Android 开发技术

许　超　主编

张晓军　赖　炜　副主编

化学工业出版社

·北京·

本书是校企合作开发的教材，以 Android 市场发布过的手机游戏作为教学案例，将案例中涉及的相关知识点，有机地融入教学过程中，并按照 Android 平台体系特征，详细介绍了各类 Android 项目开发所用的共性技术。

本书主要介绍了 Android 入门基础、Android 开发环境、Android 开发准备、Android 基本组件、Intent 和 Broadcast 应用、Android 的数据存储操作以及 Service 应用，最后通过《贪啵虎》游戏案例，综合介绍了 Android 技术的应用方法，并附有详细的开发源代码。读者通过本书的学习，将全面、系统地掌握 Android 平台相关开发技术，同时还将深入了解这些技术如何具体应用到企业开发实践中去。

本书可供高职高专计算机应用技术、计算机网络技术、软件工程、物联网等相关专业教学使用，也可供相关工程技术人员参考。

图书在版编目(CIP)数据

Android 开发技术 / 许超主编. —北京：化学工业出版社，2018.2

ISBN 978-7-122-31255-6

Ⅰ. ①A… Ⅱ. ①许… Ⅲ. ①移动终端-应用程序-程序设计 Ⅳ. ①TN929.53

中国版本图书馆 CIP 数据核字（2017）第 330451 号

责任编辑：王昕讲　　装帧设计：刘丽华

责任校对：王　静

出版发行：化学工业出版社（北京市东城区青年湖南街 13 号　邮政编码 100011）

印　　刷：三河市航远印刷有限公司

装　　订：三河市宇新装订厂

787mm×1092mm　1/16　印张 14¼　字数 364 千字　　2018 年 3 月北京第 1 版第 1 次印刷

购书咨询：010-64518888（传真：010-64519686）　售后服务：010-64518899

网　　址：http: // www. cip. com. cn

凡购买本书，如有缺损质量问题，本社销售中心负责调换。

定　　价：48.00 元

前　言

市场调查数据显示，在全球智能手机操作系统中，Android 系统占大多数，中国市场占有率更是超过了 85%，促使了就业市场对 Android 开发人员的需求量猛增。

为了更好地培养 Android 开发人才，我们与湖南东君科技实行校企深度合作，组建了课程研发团队，通过长期教学实践，逐步完成了本书的构思和编写任务。在本书编写过程中，重点体现本课程的专业基础平台课的性质，在内容的安排和深度的把握上，坚持传授 Android 应用开发技能，培养学生运用基础知识解决实际问题的能力。本书有以下几方面的特色。

（1）内容先进。本书重点介绍当前市场主流的 Android 操作系统，详细地叙述了 Android 项目开发的整个流程，引导学生了解最新的 Android 平台应用开发技术。

（2）代码丰富。本书根据高职教学实际情况，文字表达浅显易懂，案例有趣，并配有大量的开发源代码，方便教师授课，同时也便于学生理解。

（3）突出实用。本书强调实用性，以培养学生完成实际工作能力为重点，紧密联系企业实际，增加了用户需求分析、UI 设计、市场发布等内容。

（4）结构合理。本书在内容编排上紧密结合职业教育实际，符合职业院校学生的认知规律。

本书主要介绍了 Android 入门基础、Android 开发环境、Android 开发准备、Android 基本组件、Intent 和 Broadcast 应用、Android 的数据存储操作以及 Service 应用，最后通过《贪啵虎》游戏案例，综合介绍了 Android 技术的应用方法，并附有详细的开发源代码。读者通过本书的学习，将全面、系统地掌握 Android 平台相关开发技术，同时还将深入了解这些技术如何具体应用到企业开发实践中去。

本书可供高职高专计算机应用技术、计算机网络技术、软件工程、物联网等相关专业教学使用，也可供相关工程技术人员参考。

本书由许超主编，并对全书进行统稿；张晓军、赖炜担任副主编，李合军参加编写。在本书编写过程中，东君科技公司廖彦高级工程师、熊曼青高级工程师为本书编写提供了资料，刘欢、江海潮完成部分图表绘制及文档排版任务，在此对他们表示衷心的感谢。

限于编者的水平和经验，书中难免存在不妥之处，恳请读者提出批评和修改意见。

编　者

2018 年 1 月

目　录

第1章　Android入门基础

1.1　Android语言概述

1.1.1　Android的概念

Android（安卓）是一套用于移动设备的软件平台，其中包括操作系统、中间件以及一些关键应用。Android SDK 基于 Java 开发语言，提供了在 Android 平台上进行应用开发的工具和相应的 API。其应用程序由用户利用 Java 自行开发。

1.1.2　Android的发展简史

谈到 Android，首先需要了解“开发手机联盟（Open Handset Alliance）”。这是美国于 2007 年 11 月宣布组建的一个全球性联盟，这个联盟由包括中国移动、摩托罗拉、高通、宏达电子和 T-Mobile 在内的 30 多家技术和无线应用的领军企业组成。这一联盟共同开发名为 Android 的开放源代码的移动操作系统。

Android 是基于 Linux 平台的开源手机操作系统。它包括操作系统、用户界面和应用程序——移动电话工作所需的全部软件，而且不存在任何以往阻碍移动产业创新的专有权障碍。通过运营商、设备制造商、开发商和其他有关各方的合作，建立了标准化、开放式的移动电话软件平台，在移动产业内形成一个开放式的生态系统。

Android 系统采用软件堆层（software stack，又名软件叠层）的架构，主要分为 3 个部分：低层以 Linux 核心工作为基础，只提供基本功能；其他应用软件由各公司自行开发，以 Java 作为编写应用软件的基础工具。

Android 的发展历程简述如下。

1．开发手机联盟成立

2007 年 11 月 5 日，34 家公司宣布成立开发手机联盟（OHA）。

2．发布第 1 版 Android SDK

2007 年 11 月 12 日，第 1 版 Android SDK 发布。

3．Android Market 上线

2008 年 8 月 28 日，为 Android 平台手机提供软件分发和下载的 Market 正式上线，并且迅速积累了大量应用。

4．T-Mobile G1 上市

2008 年 9 月 22 日，美国移动电话运营商 T-Mobile USA 在纽约正式发布了 T-Mobile G1 手机。该款手机为台湾宏达电（HTC）代工制造，是世界上第一部使用 Android 操作系统的手机，支持 WCDMA/HSPA 网络，理论下载速率 7.2Mbps，并支持 Wi-Fi。

5．Android 1.0 SDK R1 发布

2008 年 9 月 23 日，Android 1.0 SDK R1 发布，标志着 Android 系统趋于稳定和成熟。

6．Android 宣布开放源代码

2008 年 10 月 21 日，Android 宣布开放源代码，后来又发布了 Android 1.5 SDK Release 3。

1.2 Android 的体系结构

Android 作为一个移动设备的开发平台，其软件层次结构分为操作系统（OS）、中间件（Middle-Ware）和应用程序（Application），如图 1-1 所示。中间件包括应用程序框架（Application Framework）、各种库（Libraries）和 Android 运行环境(Run-time)，简单说明如下。

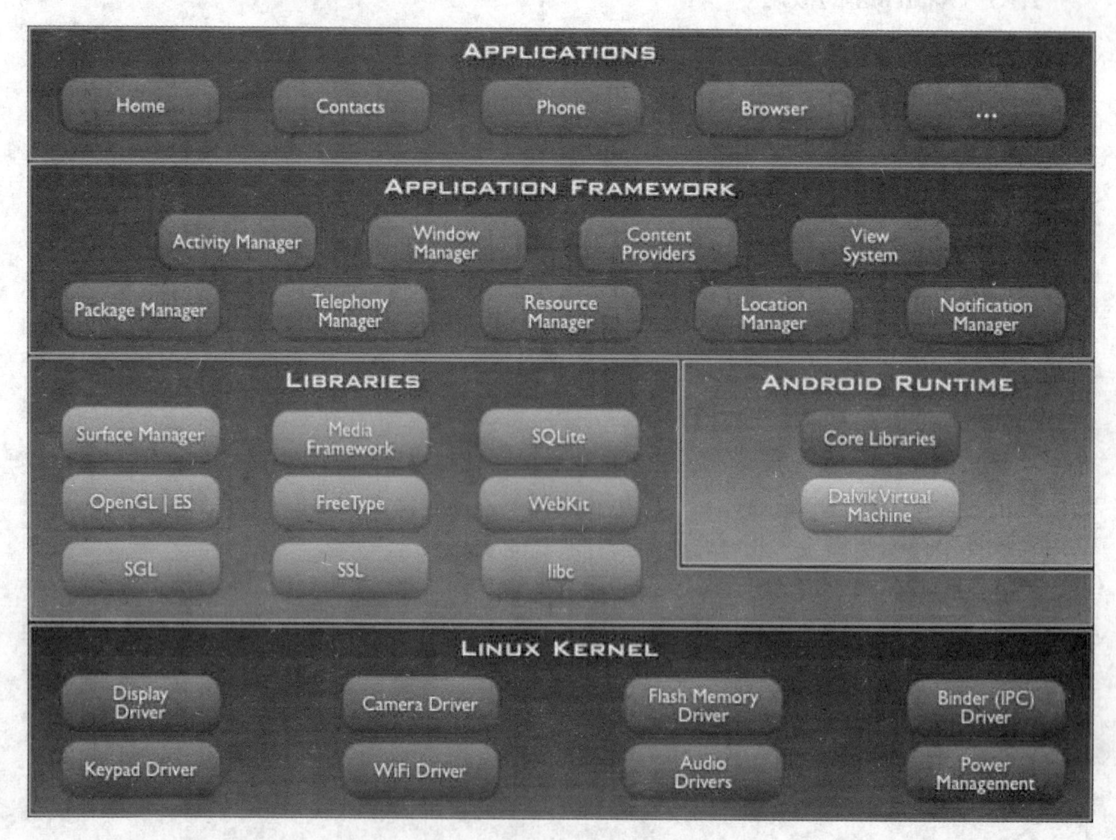

图 1-1

1.2.1 应用程序（Application）

Android 的应用程序通常涉及用户界面和用户的交互，是用户能够直接接触的部分。Android 通常将一系列核心应用程序包一起发布，包括 email 客户端、SMS 程序、日历、地图、浏览器、联系人管理程序等。Android 平台下的所用应用程序都使用 Java 语言编写，开发人员也可以使用应用程序框架的 API 开发自己的程序，显示出 Android 巨大的应用潜力。

1.2.2 应用程序框架

应用程序框架（Application Framework）设计简化了组件的重用：任何一个应用程序都可以发布它的功能，并且任何其他的应用程序都可以使用这些功能模块（应遵循框架的安全性限制）。同样地，应用程序重用机制使用户可以方便地替换程序组件。

每个应用程序可能用到的应用框架列举如下。

（1）丰富而又可扩展的视图（View）：用于构建应用程序，包括列表（list）、网格（grid）、文本框（text boxe）、按钮（button）等，以及可嵌入的 Web 浏览器。

（2）内容提供器（Context Provider）：使得应用程序可以访问另一个应用程序的数据（如联系人数据库），或者共享自己的数据。

（3）资源管理器（Resource Manager）：提供非代码资源的访问，如本地字符串、图形和布局文件（layout file）。

（4）通知管理器（Notification Manager）：使得应用程序可以在状态栏中显示自定义的提示信息。

（5）活动管理器（Activity Manager）：用来管理应用程序生命周期，并提供常用的导航返回功能。

在 Android 应用中，每个应用程序一般由多个页面组成，每个页面对应的 xml 文件对应一个 Activity。因此可以说，Android 的应用程序是由多个 Activity 的交互构成的。

1.2.3　库（Libraries）和 Android 运行环境（Run-time）

Android 系统中包含一些 C/C++库，能被不同的组件使用。它们通过 Android 应用程序框架为开发者提供服务。一些核心库简述如下。

（1）系统 C 库：一个 BSD 继承来的标准 C 系统函数库(libc)，为基于 embedded Linux 的设备定制。

（2）媒体库：基于 PacketVideo OpenCORE，支持多种常用的音频、视频格式回返和录制，同时支持静态图像文件。编码格式包括 MPEG4、H.264、MP3、AAC、AMR、JPG 和 PNG。

（3）Surface Manager：管理显示子系统，并且为多个应用程序提供 2D 和 3D 图层的无缝融合。

（4）LibWebCore：一个最新的 Web 浏览器引擎，支持 Android 浏览器和一个可嵌入的 Web 视图。

（5）SGL：底层的 2D 图形引擎。

（6）3D Libraries：基于 OpenGL ES1.0API 实现，可以使用硬件 3D 加速(如果可用)，或者使用高度优化的 3D 软加速。

（7）FreeType：位图（bitmap）和矢量（vector）字体显示。

（8）SQLite：一个对于所有应用程序可用，且功能强劲的轻型关系数据库引擎。

Android 还有一个核心库，提供 Java 编程语言核心库的大多数功能。

每一个 Android 应用程序都在它自己的进程中运行，都拥有一个独立的 Dalvik 虚拟机实例。Dalvik 被设计成一个设备，可以同时、高效地运行多个虚拟系统。Dalvik 虚拟机执行 Dalvik 可执行文件（.dex），该格式的文件针对小内存使用做了优化。同时，虚拟机基于寄存器，所有的类都经由 Java 编译器编译，然后通过 SDK 中的 dx 工具转换成.dex 格式，由虚拟机执行。

1.2.4　操作系统（OS）

Dalvik 虚拟机依赖于 Linux 内核的一些功能，比如线程机制和底层内存管理机制。Android 的核心系统服务依赖于 Linux 2.6 内核，如安全性、内存管理、进程管理、网络协议栈和驱动模型。Linux 内核同时作为硬件和软件栈之间的抽象层。

1.3 Android SDK

1.3.1 Android SDK 基础

Android SDK（Software Development Kit）提供在 Windows/Linux/Mac 平台上开发 Android 应用的组件，其中包含在 Android 平台上开发移动应用的各种工具集，不仅有 Android 模拟器和用于 Eclipse 的 Android 开发插件 ADT，而且有各种用于调试、打包和在模拟器上安装、应用的工具。

Android SDK 主要以 Java 语言为基础。通过 SDK 提供的一些工具，将其打包成 Android 平台使用的 apk 文件，然后使用 SDK 中的模拟器（Emulator）来模拟和测试该软件在 Android 平台上运行的情况和效果。

1.3.2 Android SDK 目录结构

Android SDK 1.1 的目录结构如图 1-2 所示。

Android SDK 1.5 的目录结构如图 1-3 所示。

```
├─android.jar
├─documentation.html
├─docs
├─samples
│    ├─ApiDemos
│    ├─HelloActivity
│    ├─Home
│    ├─LunarLander
│    ├─NotePad
│    ├─SkeletonApp
├─tools
```

图 1-2

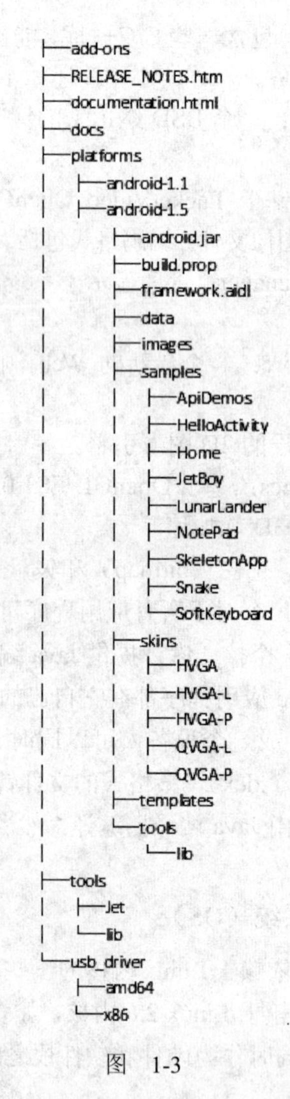

图 1-3

（1）Add-ons 目录下的 google_apis3 提供的 API 包，主要包括 API Documentation.html 和 docs 目录下的文档

（2）RELEASE_NOTES.html 是 SDK 的发布说明。

（3）usb_driver 目录下包含 amd64 和 x86 的驱动文件。

（4）Tools 目录下包含一些通用的工具文件。

（5）在 Platforms 目录下，针对每个版本的 SDK，提供了对应的 API 包及其示例。

① android.jar 是包含全部 API 的压缩包；

② samples 目录下是 SDK 附带的一些示例；

③ skins 目录下是其支持的几种外观像素；

④ templates 目录下是一些常用的文件模板；

⑤ tools 目录下是一些常用的辅助工具。

1.3.3 Android. jar 及内部结构

Android.jar 是一个标准 jar 压缩包，其内部是编译后的 class 文件，包含全部的 API，结构如图 1-4 所示。

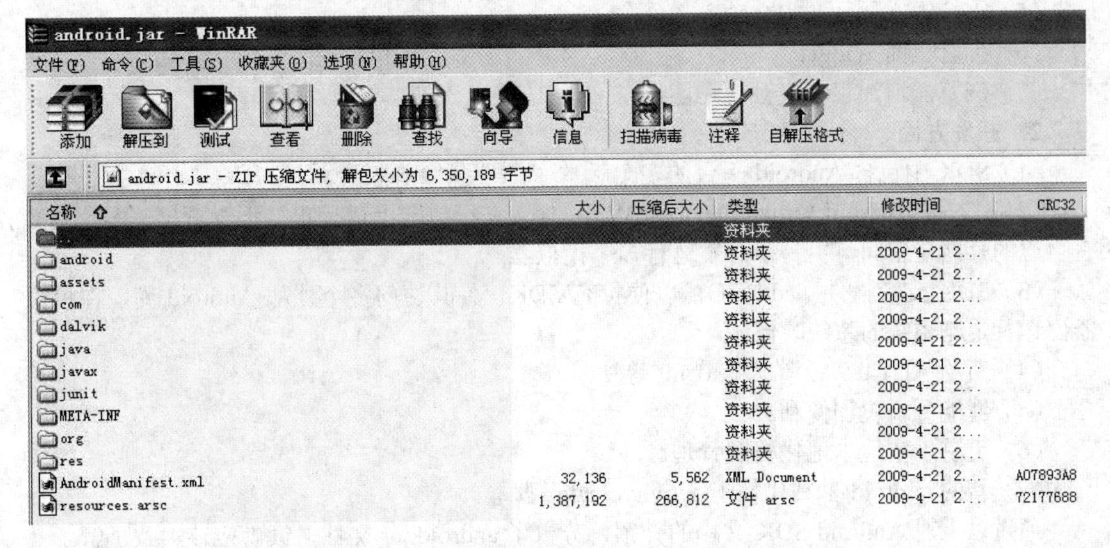

图 1-4

通过图 1-4 可以了解其模块的划分和结构，有助于用户阅读和查找 SDK 文档。

1.3.4 Android API 核心开发包

SDK 参考文档是按照包结构组织的，以便用户很清晰地看到 API 的结构。其核心包中的模块主要包括以下内容。

（1）android.app：提供高层的程序模型，提供基本的运行环境。

（2）android.content：包含各种访问和发布设备上的数据的类。

（3）android.database：通过内容提供者浏览和操作数据库，包含底层 API 处理数据库。

（4）android.graphics：底层图形库，包含画布，通过过滤、点、矩形，将图形直接绘制到屏幕上，并作为核心渲染包，提供图形渲染功能。

（5）android.location：定位和提供相关服务的类。

（6）android.media：提供外挂多种音频、视频等媒体接口的类。

（7）android.net：通过的 java.net.*接口，提供帮助网络访问的类。

（8）android.os：提供系统服务、消息传输和进程间通信（IPC）。

（9）android.provider：提供访问 Android 内容提供者的类。

（10）android.telephony：提供与拨打电话相关的 API 交互。

（11）android.view：提供基础的用户界面接口框架。

（12）android.util：涉及工具性的方法，例如时间、日期的操作。

（13）android.widget：包含各种 UI 元素（大部分是可见的），在应用程序的屏幕中使用。

（14）android.webkit：包含一系列基于 Web 内容的 API。

1.3.5 Android SDK 1.5 的新特性

1．系统方面

（1）采用当时最新的 Linux 内核 2.6.27 版本。

（2）精简了用户界面。

（3）拥有全新的视频录制功能，可以上传视频到 Youtube，上传照片到 Picasa。

（4）支持软键盘。

（5）支持中文显示和中文输入。

（6）拥有桌面 Widget。

（7）浏览器增加了多点触摸功能。

2．开发方面

（1）SDK 中包含 Android 平台的多个版本（1.1 版和 1.5 版）。

（2）引入了 Android Virtual Devices(AVD)，使之在模拟器上运行更接近于真机。每个 AVD 有自己的存储卡空间，更易于开发运行多个模拟器。

（3）SDK 支持插件（add-on），以便扩充 SDK，使其访问多个外部 Android 库，在模拟器内运行定制系统的系统映像。

（4）新的 ADT 0.9 版增加了 JUnit 等功能。

（5）性能分析更加便利。

（6）更易于实现本地化资源管理。

（7）新的 android 工具代替了 activitycreator 脚本。

通过以上对 Android SDK 文档的介绍，分析了 android.jar 文件，读者应该大致了解了其内部 API 的结构和组织方式。如果想深入了解各个文件包含的 API 及其用法，必须学会阅读和查找 SDK 文档。读者使用浏览器打开 SDK 目录下的 documentation.html 文件，可详细阅读其内容。

第 2 章　Android 开发环境

2.1　Android 开发环境搭建

第 1 章介绍了 Android 的基础知识，本章详细介绍搭建 Android 开发环境的步骤。

2.1.1　Android 开发系统要求

Android 开发系统要求如下：

（1）Windows XP（32 位）或 Vista（32 或 64 位）。

（2）Mac OS X 10.4.8 或更高版本（仅对 x86 版）。

（3）Linux（在 Linux Ubuntu Dapper Drake 上经过测试）。

2.1.2　下载所需软件包

在 http://java.sun.com/javase/downloads/index.jsp 下载 JDK，如图 2-1 所示。

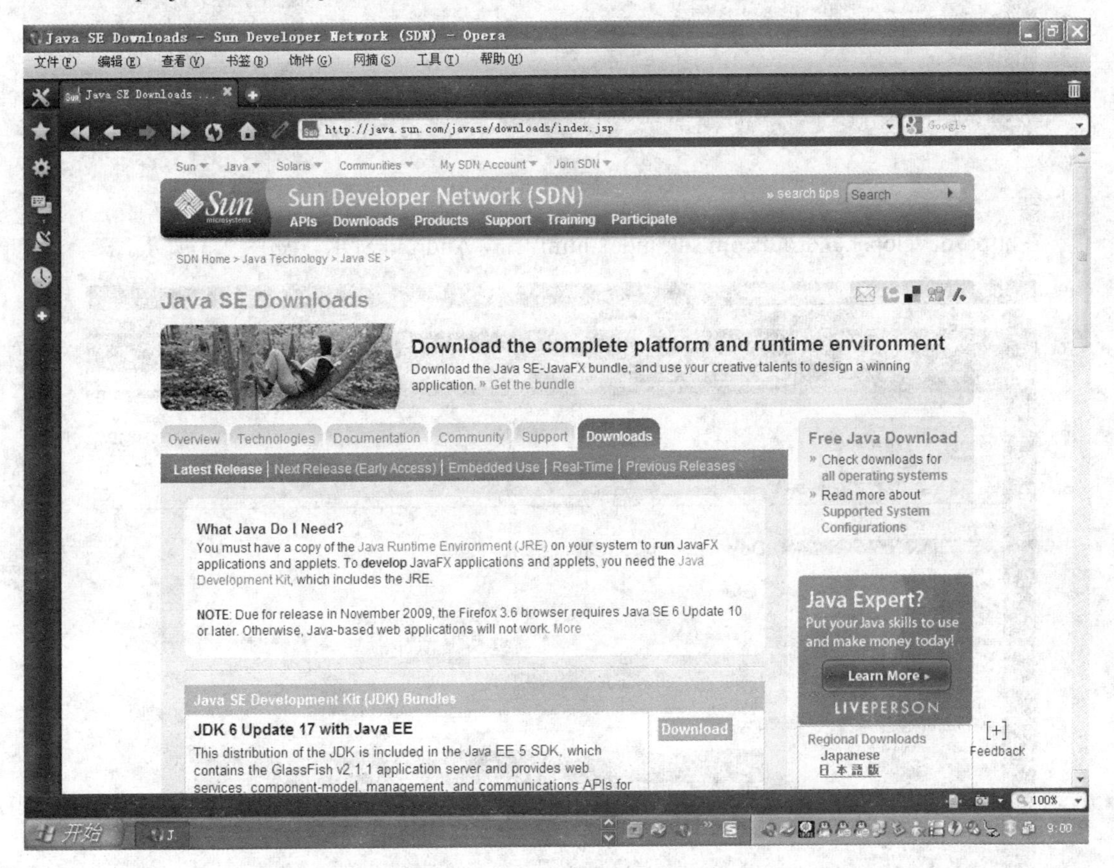

图　2-1

在http://www.eclipse.org/downloads/下载 EcLipse，如图 2-2 所示。

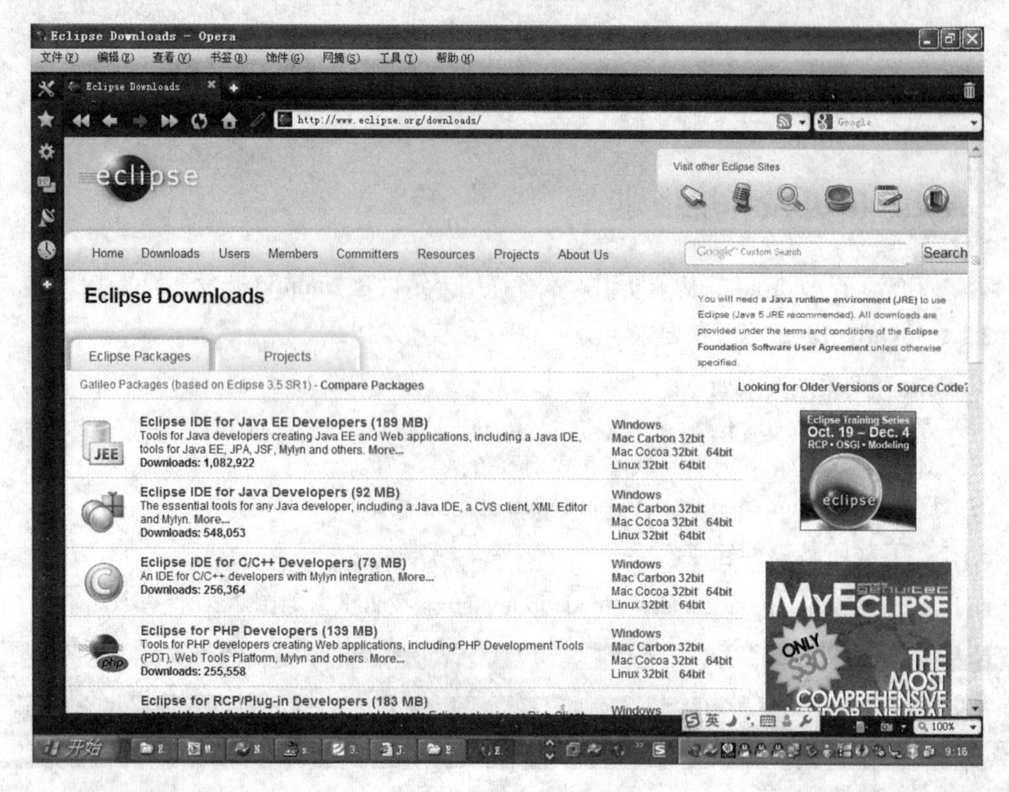

图 2-2

在http://developer.android.com/sdk/index.html 下载 Android SDK，如图 2-3 所示。

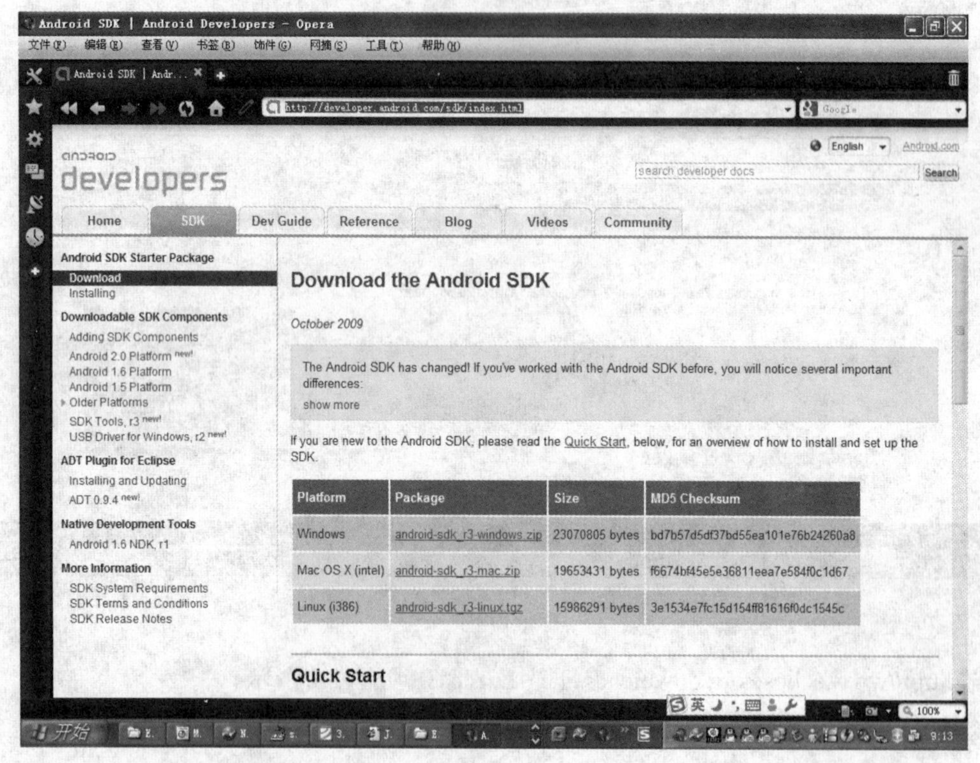

图 2-3

2.1.3　安装 Android SDK

Android SDK 同 Eclipse 一样，直接解压缩就可以打开。假定下载的压缩包解压缩到文件夹 F:\Android 中，如图 2-4 所示。

图　2-4

将 Android SDK 中 tools 目录的绝对路径添加到系统 PATH 中。然后，打开“系统属性”对话框，单击“环境变量”按钮，如图 2-5 所示。

在“环境变量”对话框中，将 PATH 的值设置为 SDK 中 tools 的绝对路径，如图 2-6 所示。

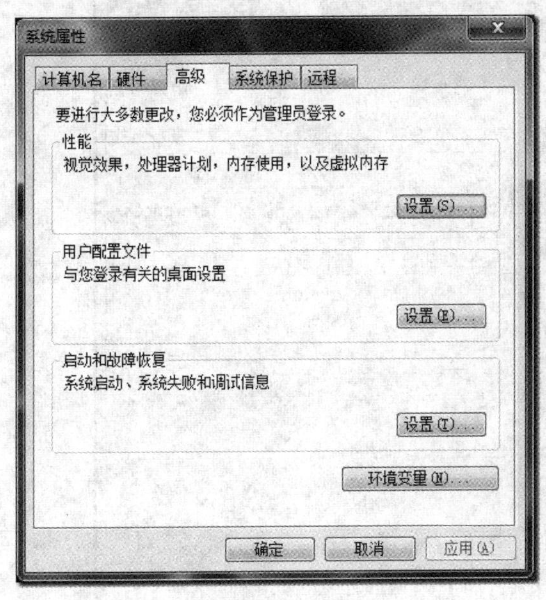

图　2-5

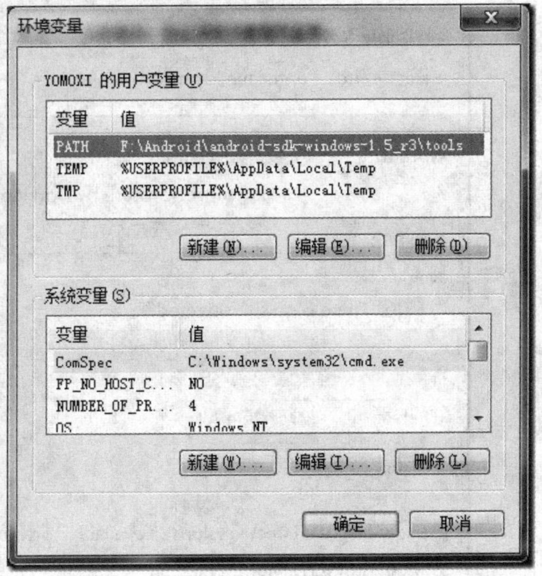

图　2-6

单击“确定”按钮，重新启动计算机，进入 cmd 命令窗口，检查 SDK 是否安装成功。

运行 android –h，如果输出如图 2-7 所示，表明安装成功。

2.1.4　安装 ADT

（1）启动 Eclipse，然后单击菜单中的 Help→Install New Software，弹出 AvaiLable Software 对话框，如图 2-8 所示。

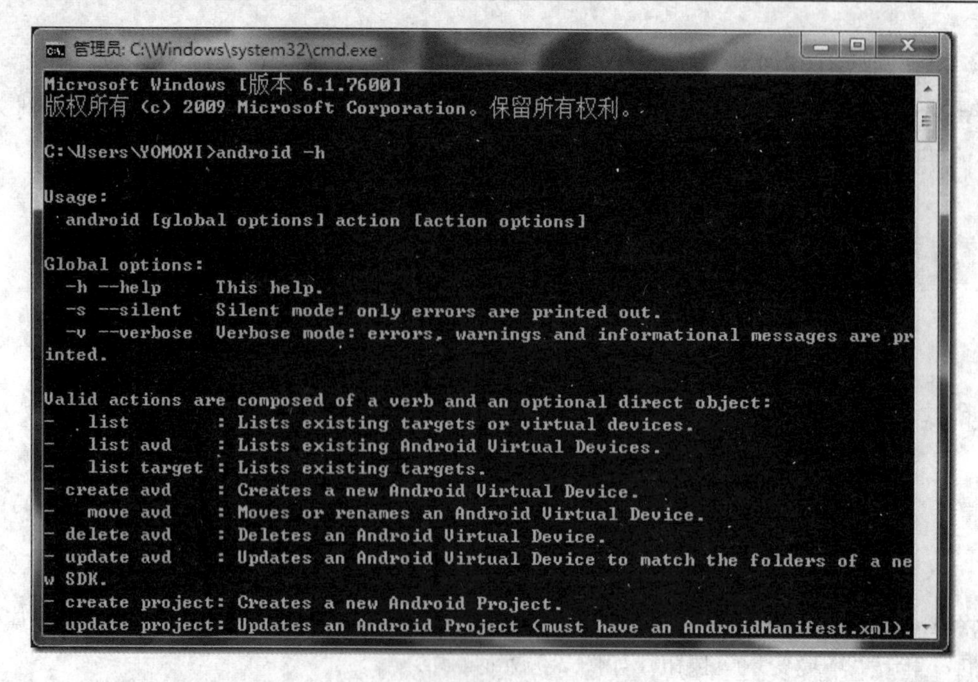
管理员: C:\Windows\system32\cmd.exe
Microsoft Windows [版本 6.1.7600]
版权所有 (c) 2009 Microsoft Corporation。保留所有权利。
C:\Users\YOMOXI>android -h
Usage:
android [global options] action [action options]
Global options:
-h --help This help.
-s --silent Silent mode: only errors are printed out.
-v --verbose Verbose mode: errors, warnings and informational messages are printed.
Valid actions are composed of a verb and an optional direct object:
- list : Lists existing targets or virtual devices.
- list avd : Lists existing Android Virtual Devices.
- list target : Lists existing targets.
- create avd : Creates a new Android Virtual Device.
- move avd : Moves or renames an Android Virtual Device.
- delete avd : Deletes an Android Virtual Device.
- update avd : Updates an Android Virtual Device to match the folders of a new SDK.
- create project: Creates a new Android Project.
- update project: Updates an Android Project (must have an AndroidManifest.xml).

图 2-7

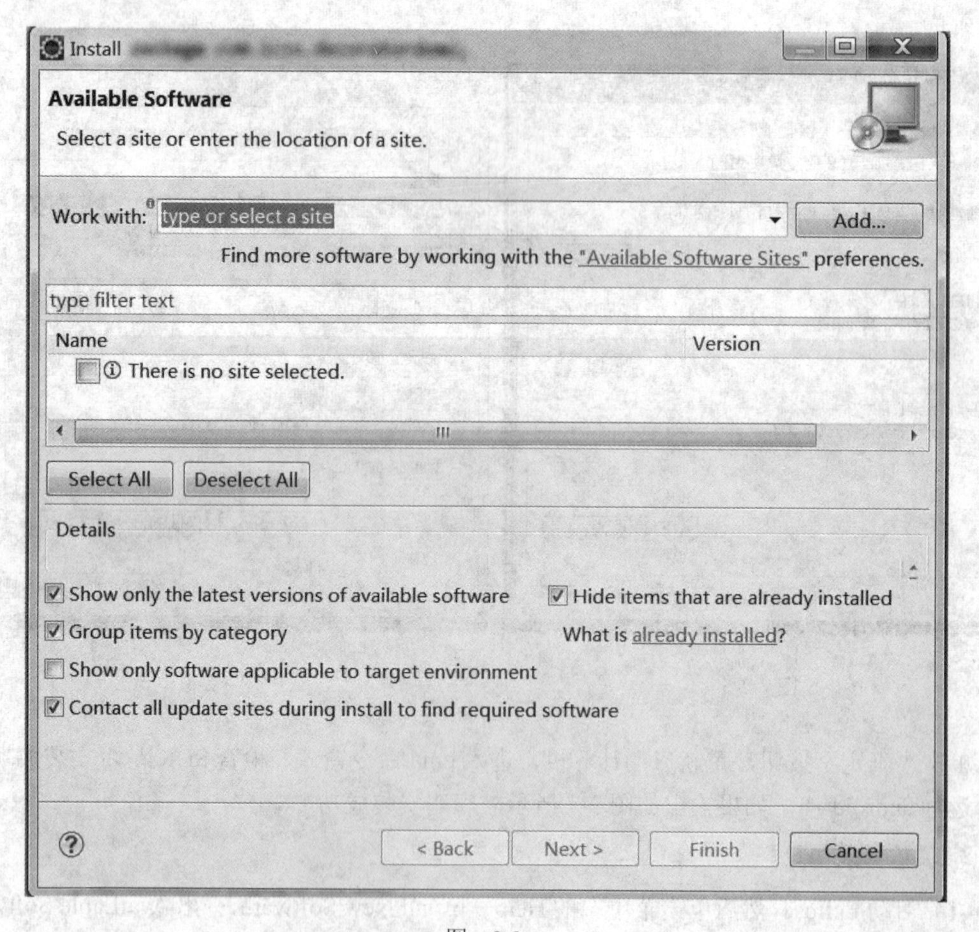
Install
Available Software
Select a site or enter the location of a site.
Work with: type or select a site
Add...
Find more software by working with the "Available Software Sites" preferences.
type filter text
Name
Version
There is no site selected.
Select All
Deselect All
Details
Show only the latest versions of available software
Hide items that are already installed
Group items by category
What is already installed?
Show only software applicable to target environment
Contact all update sites during install to find required software
< Back
Next >
Finish
Cancel

图 2-8

（2）单击 Add...按钮，弹出 Add Repository 对话框，如图 2-9 所示。

图　2-9

（3）在 Name 框中输入 Android Plugin；在 Location 框中输入对应的 URL：https://dl-ssl.google.com/android/eclipse/，然后单击 OK，按钮，弹出如图 2-10 所示对话框。

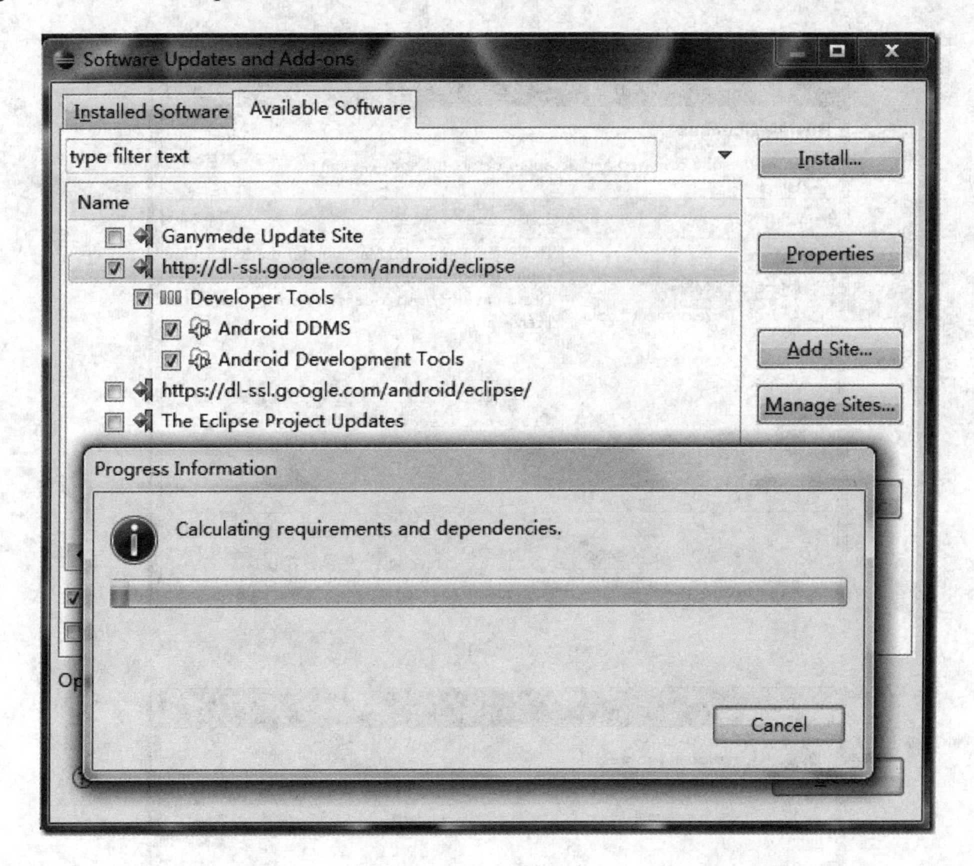

图　2-10

（4）勾选 Android DDMS 和 Android Development Tools，然后单击 Next 按钮，开始安装，如图 2-11 所示。

（5）单击 Next 按钮，并且选择接受相关的许可，执行安装，如图 2-12～图 2-14 所示。

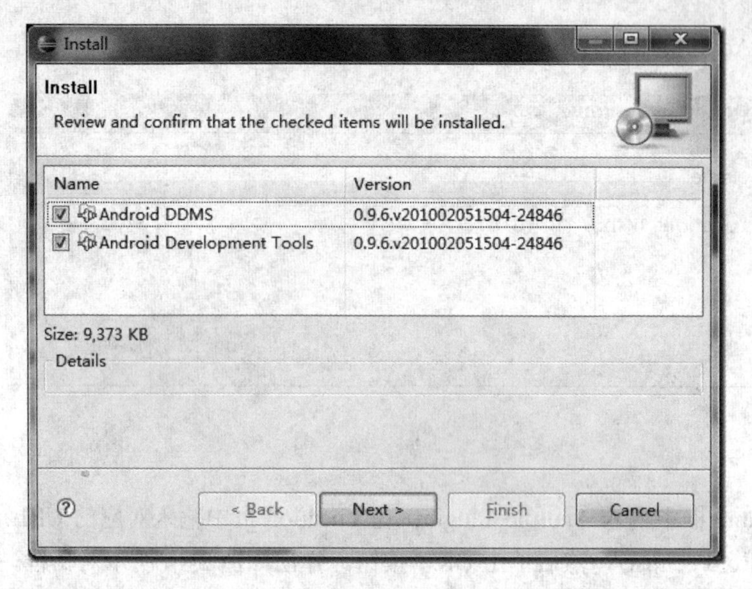

图 2-11

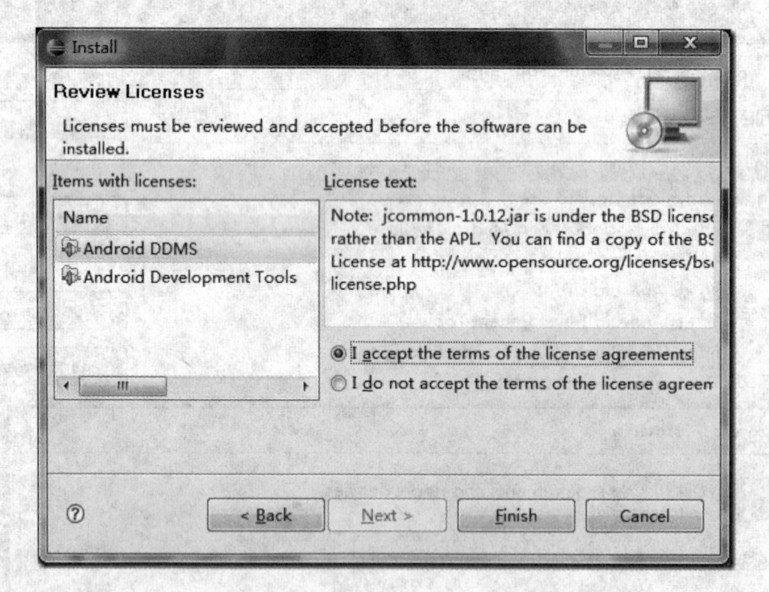

图 2-12

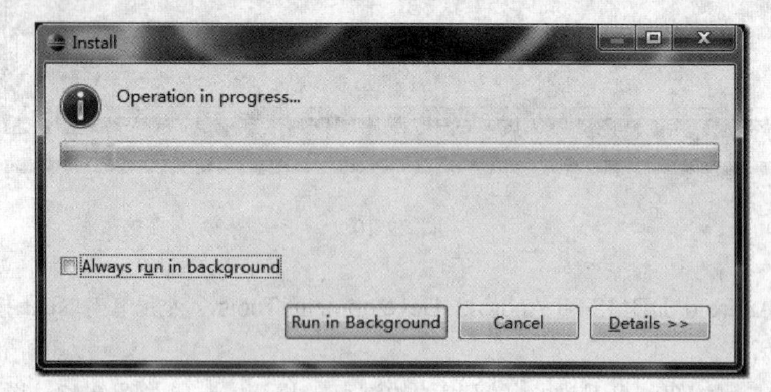

图 2-13

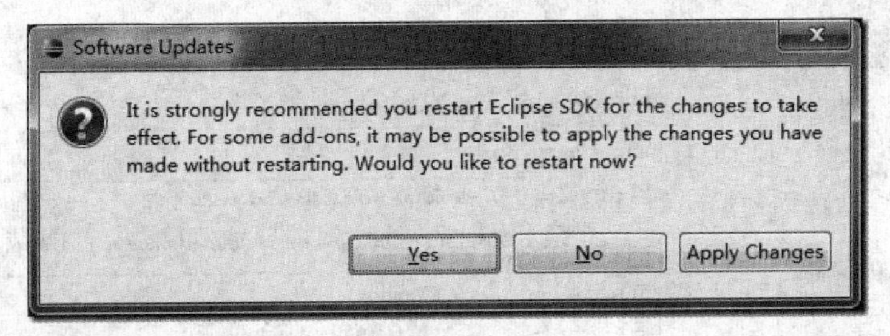

图　2-14

（6）单击 Yes 按钮，重启 Eclipse，完成安装。

2.1.5　设置 SDK

打开 Eclipse IDE，然后在菜单中选择 Window→Preferences，打开 Preferences 窗口，并选中 Android，如图 2-15 所示。

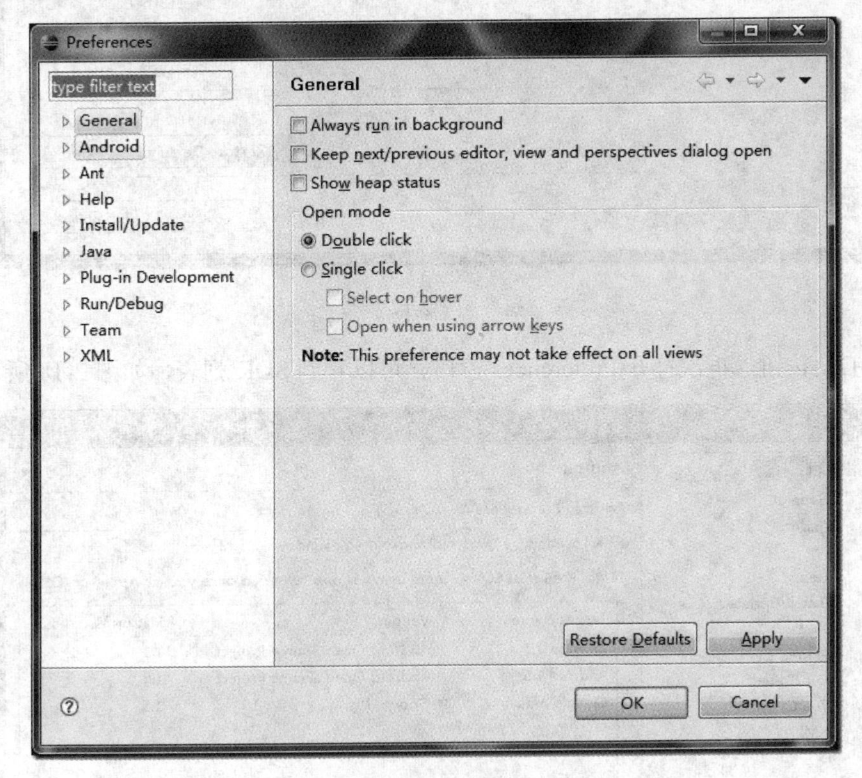

图　2-15

忽略弹出的错误窗口，如图 2-16 所示，直接设定 SDK Location 为 SDK 的安装目录，如图 2-17 所示。

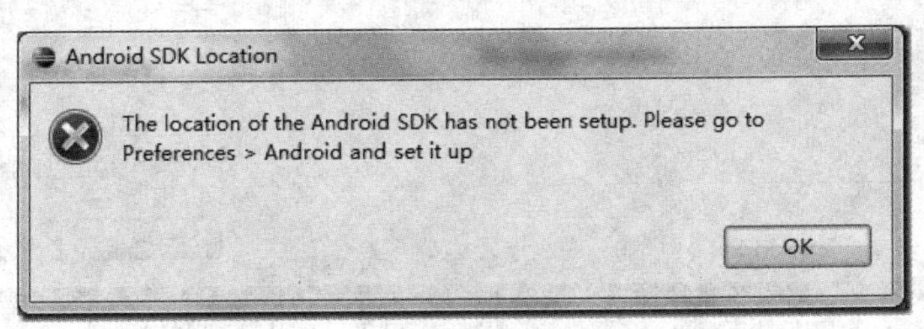

图　2-16

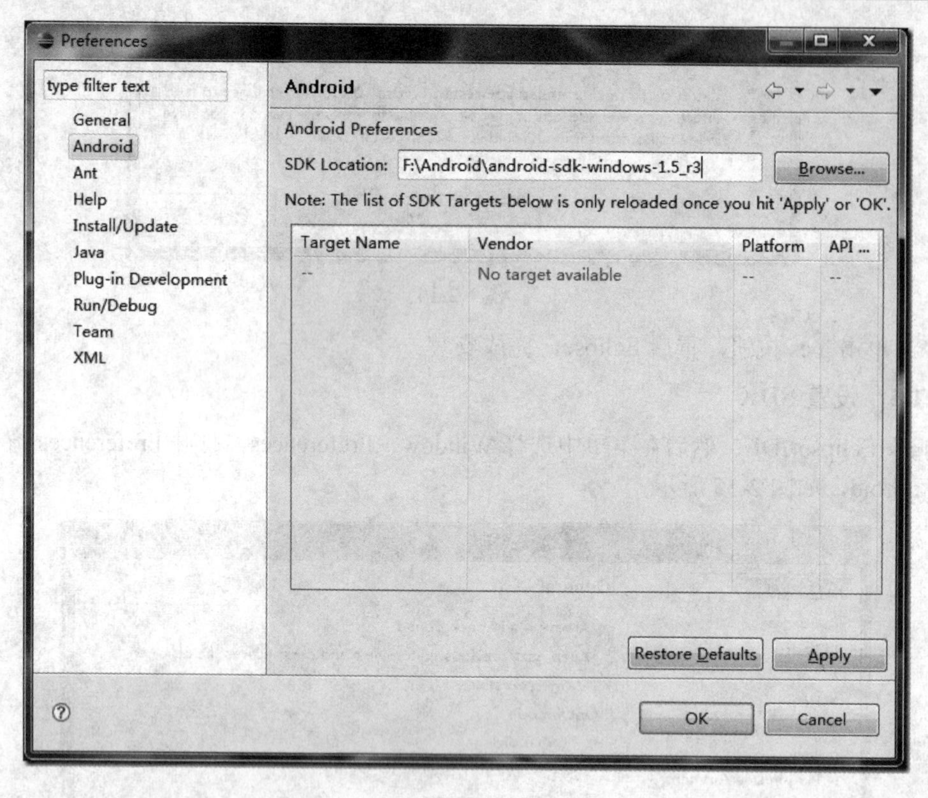

图 2-17

单击 OK 按钮，再次打开 Preferences 窗口，可以看到 SDK 列表，如图 2-18 所示。

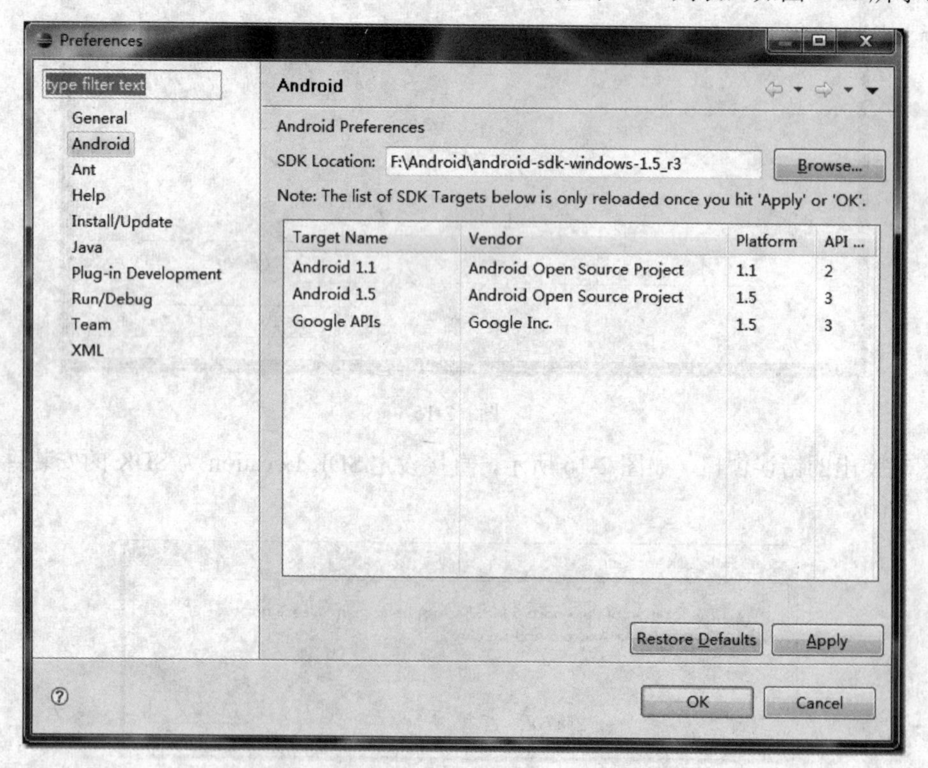

图 2-18

2.1.6　验证开发环境

在 Eclipse IDE 菜单中单击 File→New→Project，如图 2-19 所示，弹出 New Project 对话框，如图 2-20 所示。

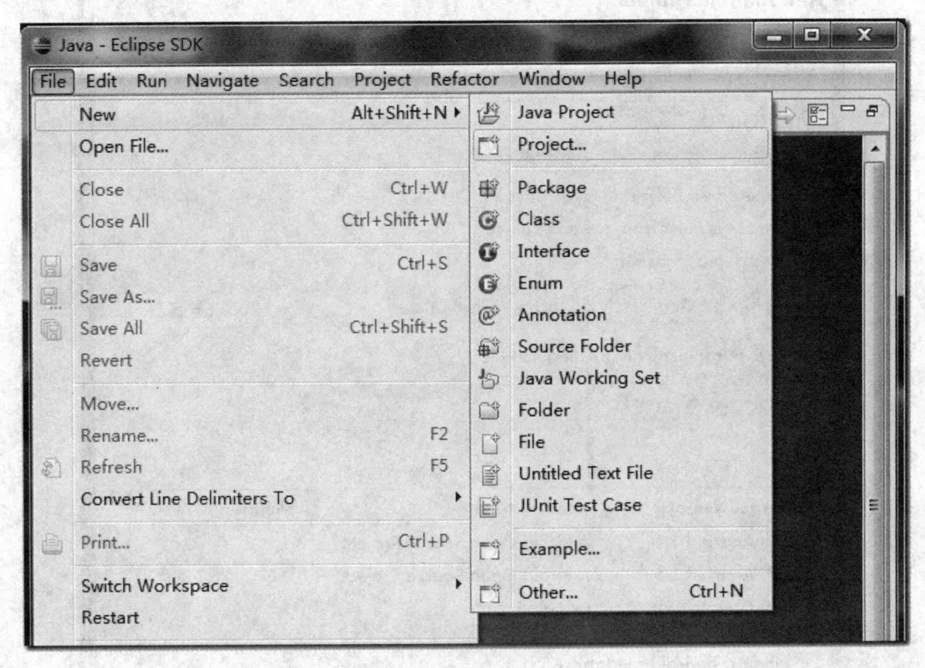

图　2-19

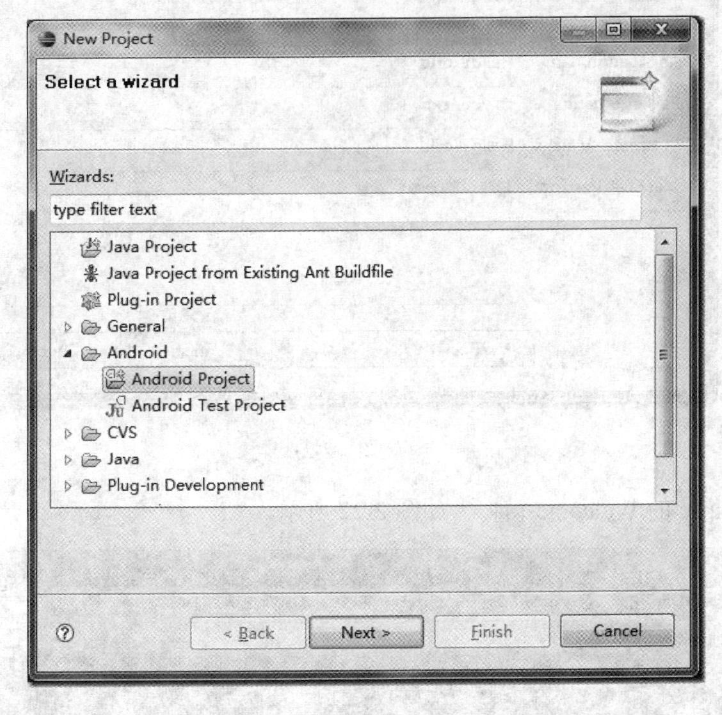

图　2-20

选择 Android Project，然后单击 Next 按钮，弹出 New Android Project 对话框，如图 2-21 所示。

如图 2-21 所示设置相关参数，然后单击 Finish 按钮。

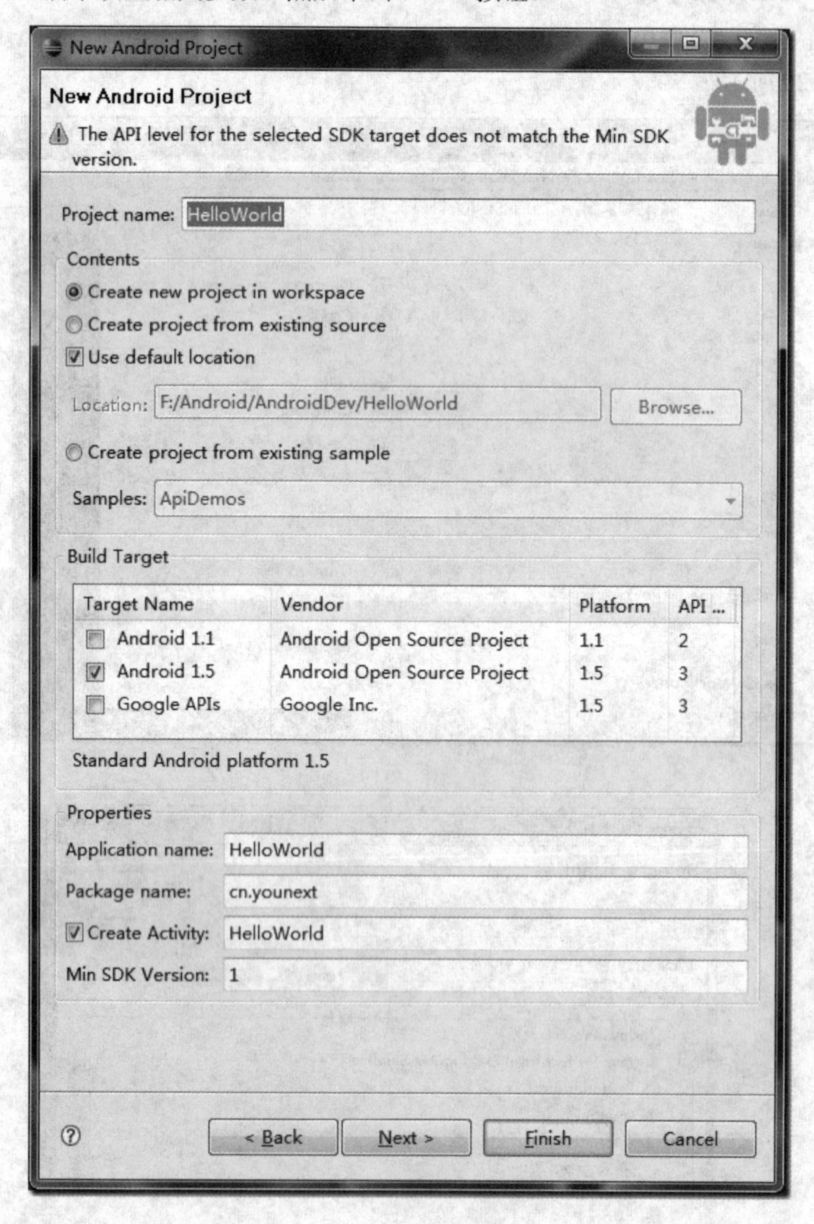

图 2-21

关闭 Eclipse 的 Welcome 窗口，如图 2-22 所示。

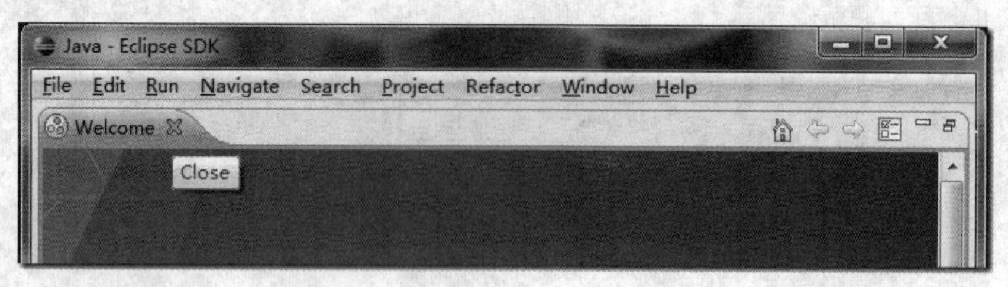

图 2-22

创建的项目如图 2-23 所示。

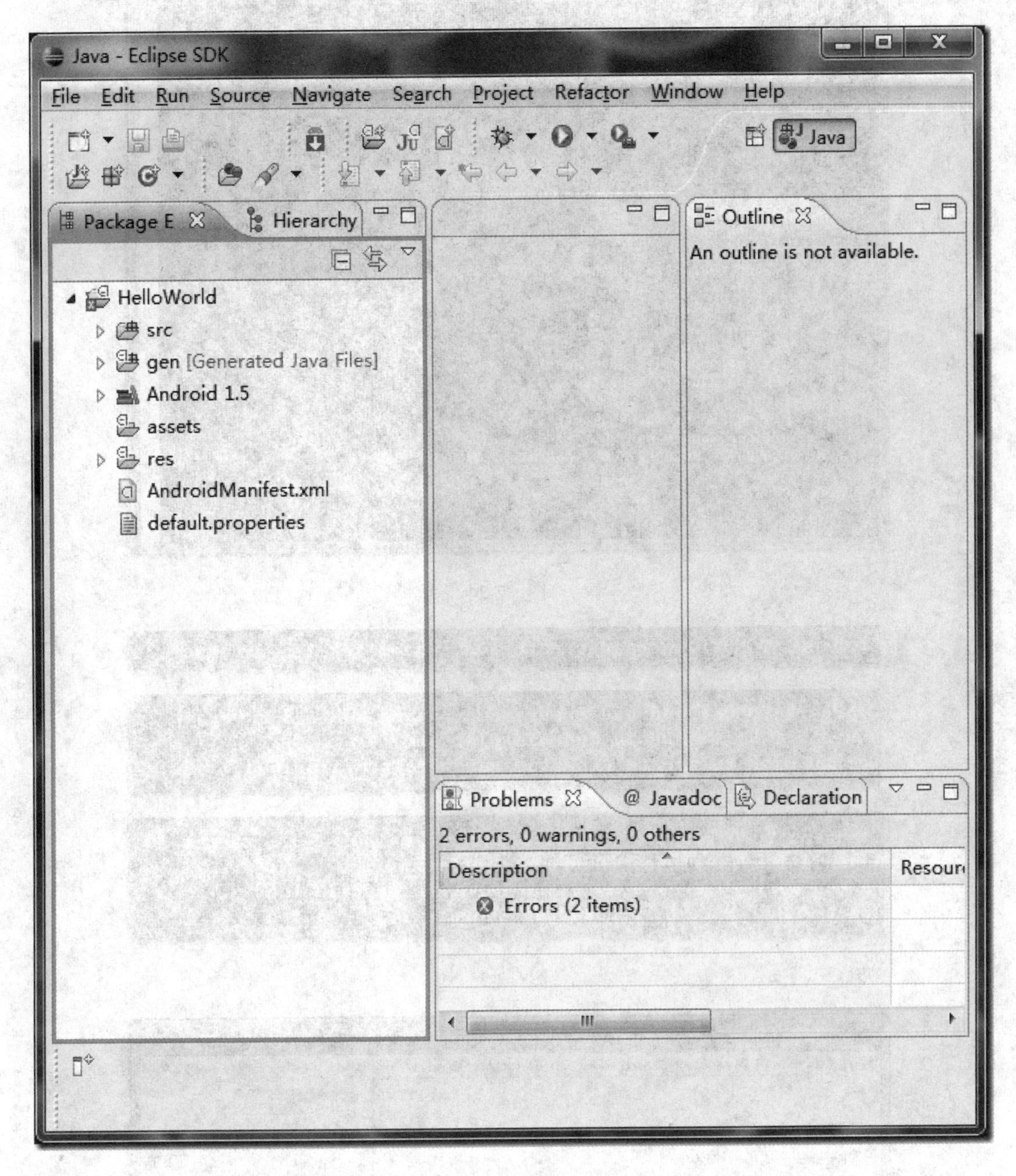

图　2-23

接下来，创建 Android 虚拟设备 AVD。打开 cmd 控制台，执行 android list target 命令，查看可用的平台，如图 2-24 所示。然后，根据 android create avd –name <AVD 名字> –target <id> 格式创建 AVD，如图 2-25 所示，完成自定义的 Android Virtual Device。

最后，配置 Eclipse 的 Run Configurations。在 Eclipse SDK 菜单中单击 Run→Run Configurations，如图 2-26 所示。

```
管理员: C:\Windows\system32\cmd.exe
Microsoft Windows [版本 6.1.7600]
版权所有 (c) 2009 Microsoft Corporation。保留所有权利。

C:\Users\YOMOXI>android list target
Available Android targets:
id: 1
     Name: Android 1.1
     Type: Platform
     API level: 2
     Skins: HVGA (default), HVGA-L, HVGA-P, QVGA-L, QVGA-P
id: 2
     Name: Android 1.5
     Type: Platform
     API level: 3
     Skins: HVGA (default), HVGA-L, HVGA-P, QVGA-L, QVGA-P
id: 3
     Name: Google APIs
     Type: Add-On
     Vendor: Google Inc.
     Description: Android + Google APIs
     Based on Android 1.5 (API level 3)
     Libraries:
      * com.google.android.maps (maps.jar)
          API for Google Maps
     Skins: QVGA-P, HVGA-L, HVGA (default), QVGA-L, HVGA-P

C:\Users\YOMOXI>
```

图 2-24

```
C:\Users\YOMOXI>android create avd --name sdk_1_5 --target 2
```

```
C:\Users\YOMOXI>android create avd --name sdk_1_5 --target 2
Android 1.5 is a basic Android platform.
Do you wish to create a custom hardware profile [no]
```

```
C:\Users\YOMOXI>android create avd --name sdk_1_5 --target 2
Android 1.5 is a basic Android platform.
Do you wish to create a custom hardware profile [no]
Created AVD 'sdk_1_5' based on Android 1.5
```

图 2-25

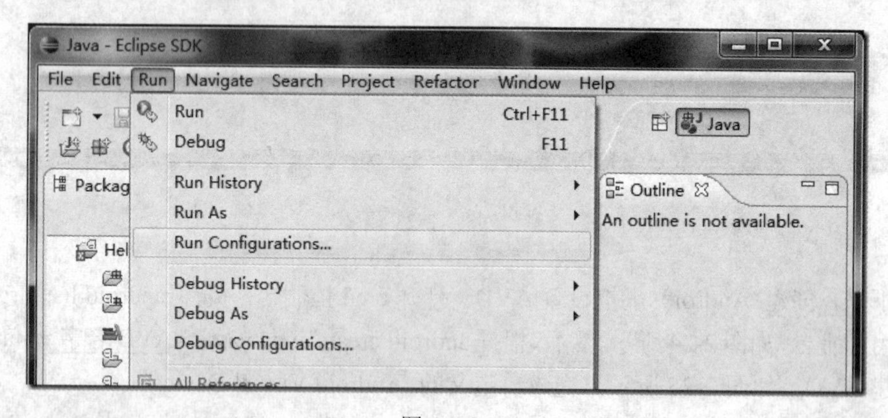

图 2-26

在弹出的窗口中双击 Android Application，创建一个新的配置文件，并设置 Name 项，如图 2-27 所示。

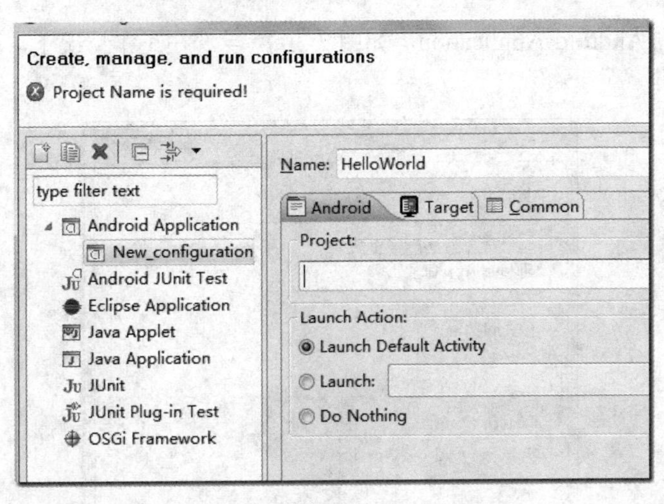

图　2-27

指定右侧 Android 选项卡中的 Project 项目，并且在右侧 Target 选项卡中勾选创建的 AVD，然后单击 Apply 按钮，最后单击 Run 按钮，如图 2-28 和图 2-29 所示。

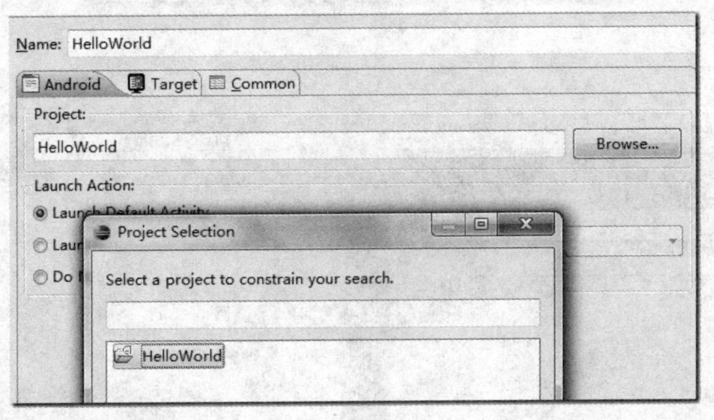

图　2-28

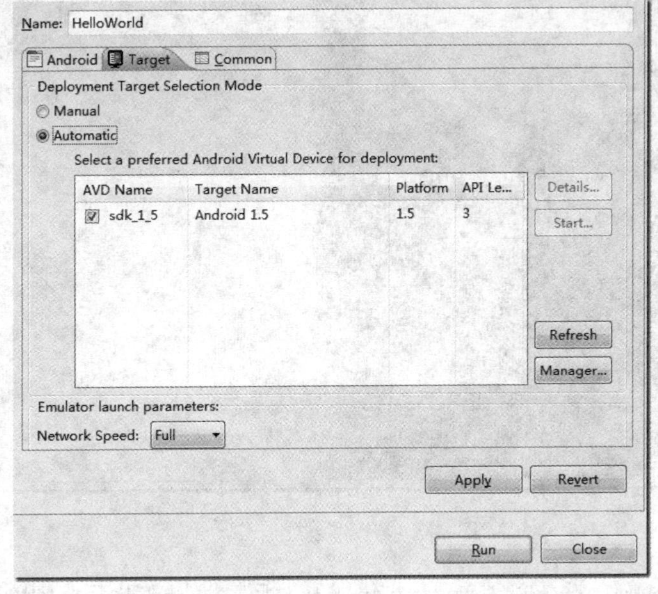

图　2-29

选择启动方式 Android Application，如图 2-30 所示。在正常情况下，可以看到模拟器界面，如图 2-31 所示。

图 2-30

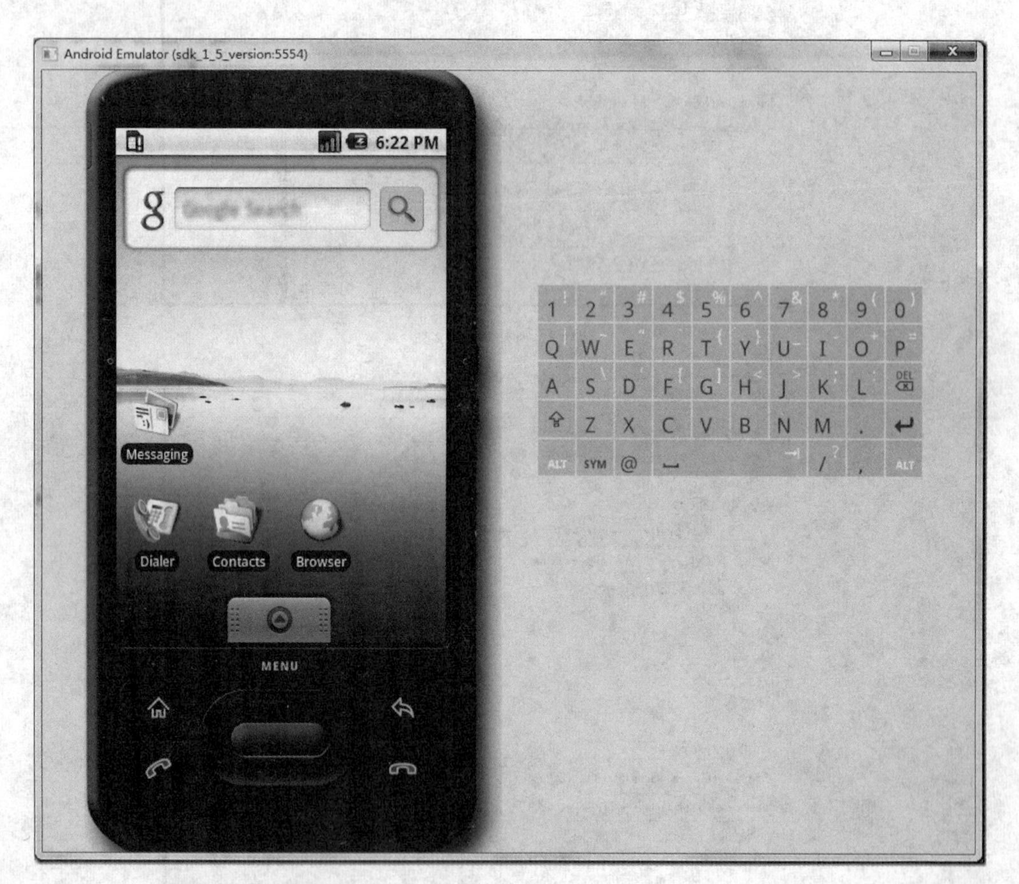

图 2-31

通过执行上述操作，搭建起 Android 集成开发环境可以在此环境中创建 Android 项目。

对于使用 Eclipse 开发的 Android 的应用程序，是在模拟器中进行调试的。下面介绍 Android 模拟器。

2.2　Android 模拟器

2.2.1　模拟器概述

模拟器如图 2-31 所示，它由两个部分组成：左边部分模拟手机显示，右边部分模拟手机键盘输入。模拟器手机部分还内置了一些自带的程序，用于打电话、发短信等。

2.2.2　使用命令行工具管理模拟器

可以使用模拟器管理工具来管理模拟器。SDK 中提供了一个 Android 命令行工具（在 Android-sdk/tools 中），用于创建新项目或是管理模拟器。2.1 节创建了一个模拟器，创默认将在 C:\Documents and Settings\Administrator\.android\avd\目录下生成对应的.avd 文件。

使用命令行工具提供的 android list avd 命令列出所有的模拟器，如图 2-32 所示。

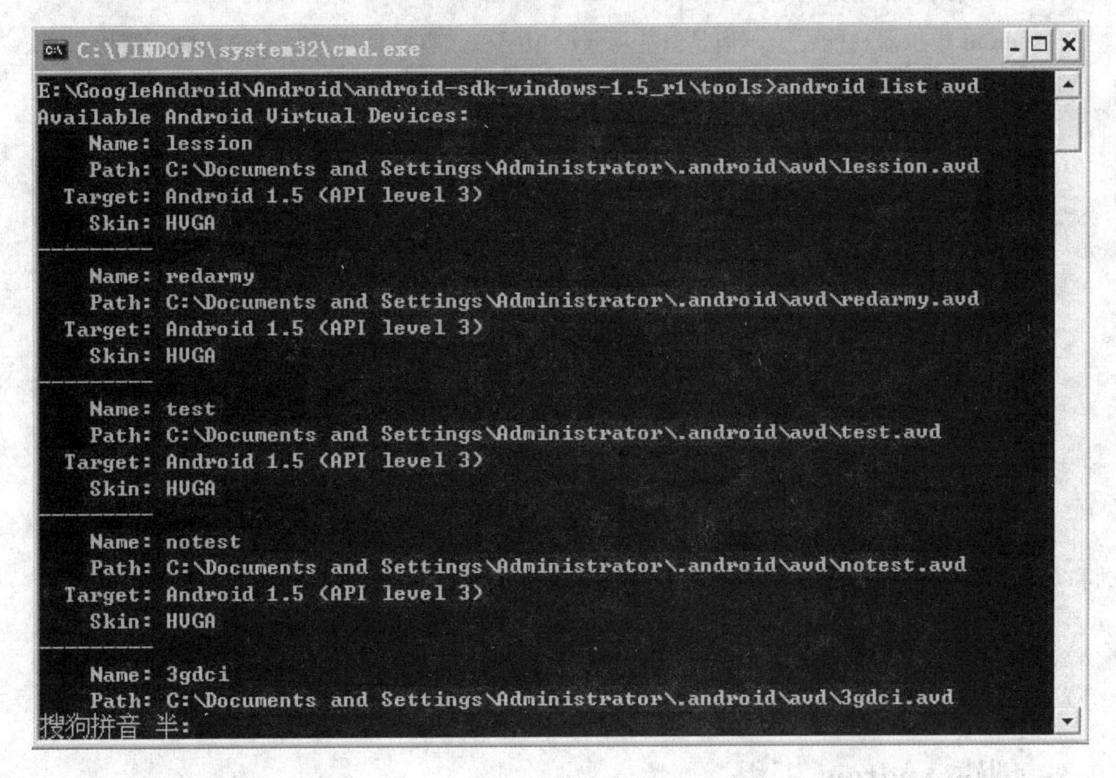

图　2-32

2.2.3　操作模拟器

模拟器是 Android 应用程序开发者最常用的工具，它提供了很多功能值得用户多做尝试。

1．切换模拟器布局

在命令行运行 android list targets 命令后，屏幕上将列出所有支持的模拟器类型。以第二种类型（id 2）模拟器为例，列出信息如图 2-33 所示。

```
id: 2
     Name: Android 1.5
     Type: Platform
     API level: 3
     Skins: HVGA (default), HVGA-L, HVGA-P, QVGA-L, QVGA-P
```

图 2-33

图中，Skins 字段中列出所有支持的模拟器布局。默认有 HVGA（分辨率 480×320）与 QVGA（分辨率 320×240)两种画面配置选项可供选择。HVGA 与 QVGA 可以再各自分为-L（landscape，横排）与-P（portrait，竖排）。

要创建 QVGA 模式的模拟器，在 android create avd 命令后，加上-skin QVGA 选项。若要将默认的 HVGA 竖排显示改为横排，使用快捷键，直接切换屏幕。

2．切换屏幕

在 Windows 操作系统中按 Ctrl+F12 键，或是在 Mac OS X 操作系统中按 fn+7 键，Android 模拟器的屏幕就从默认的直式显示切换成横式显示。同样地，可以切换过来。

2.2.4 模拟器与真机的区别

Android 模拟器功能强大，但是只能尽量模拟手机，有些功能还是无法模拟。例如：

（1）模拟器不支持呼叫和接听实际来电，但可以通过控制台模拟电话呼叫（呼入和呼出）。

（2）模拟器不支持 USB 连接。

（3）模拟器不支持相机/视频捕捉。

（4）模拟器不支持音频输入（捕捉），但支持输出（重放）。

（5）模拟器不支持扩展耳机。

（6）模拟器不支持蓝牙。

（7）模拟器不能确定连接状态。

（8）模拟器不能确定电池电量水平和充电状态。

（9）模拟器不能确定 SDK 卡的插入/弹出。

2.2.5 使用模拟器的注意事项

（1）平时使用模拟器 Emulator 测试开发时，若计算机提示“系统 C 盘空间不足”，表示 Android 模拟器运行时生成几个以 tmp 为后缀名的临时文件，可能占用了几 GB 磁盘空间。可以到 C:\Documents and Settings\用户名\Local Settings\Temp\AndroidEmulator 目录清理。

（2）在使用 Eclipse 开发工具进行调试时，第一次运行程序启动模拟器的时间比较长，大概需要 1 分钟。为此，启动模拟器后，每次运行新的程序时不要关闭旧的模拟器，直接在 Eclipse 开发工具里单击“运行”即可。

2.3 创建 Android 工程

2.1 节和 2.2 节创建了使用 Eclipse 来开发 Android 应用程序的集成环境和运行 Android 应用程序的虚拟设备。下面介绍创建 Android 应用程序的操作步骤。

2.3.1 创建 HelloAndroid 项目

（1）启动 Eclipse，选择 File→Project→Android→Android Project，然后单击 Next 按钮，如图 2-34 所示。

（2）填写项目信息，如图 2-35 所示。

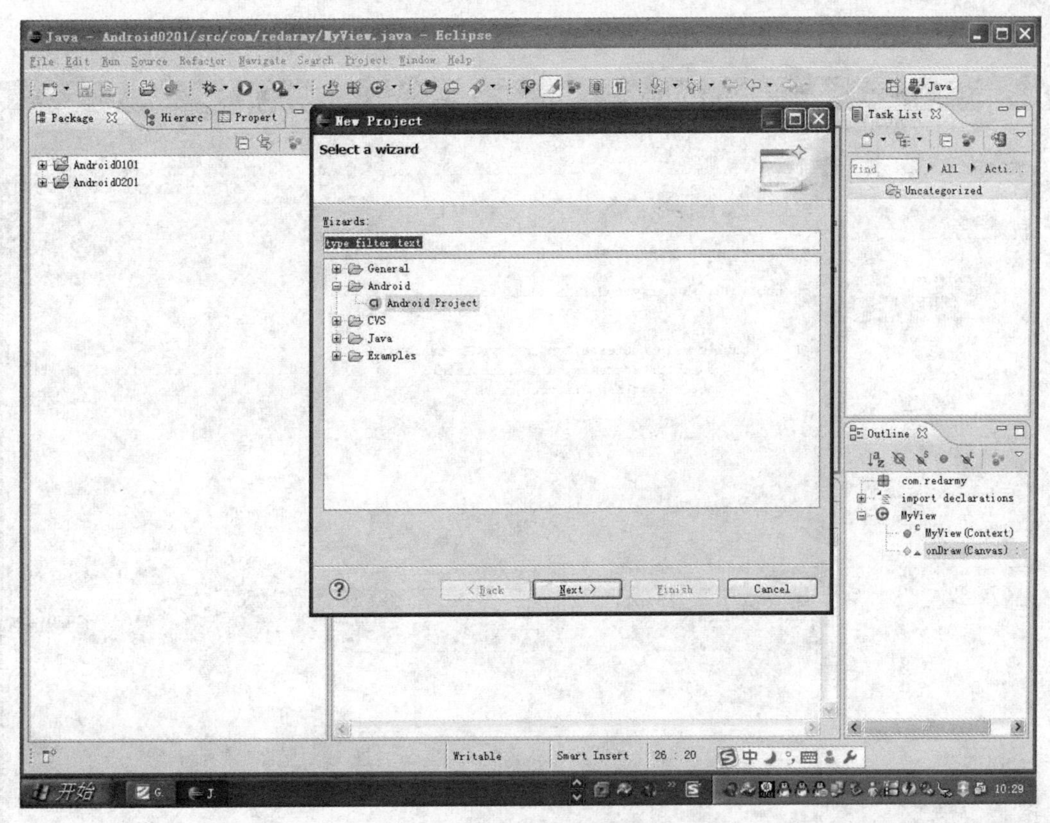
Java - Android0201/src/com/redarmy/MyView.java - Eclipse
File Edit Run Source Refactor Navigate Search Project Window Help
Package
Hierarc
Propert
Android0101
Android0201
New Project
Select a wizard
Wizards:
type filter text
General
Android
Android Project
CVS
Java
Examples
< Back
Next >
Finish
Cancel
Task List
Find
All
Acti...
Uncategorized
Outline
com.redarmy
import declarations
MyView
MyView(Context)
onDraw(Canvas)
Writable
Smart Insert
26 : 20
开始
10:29

图　2-34

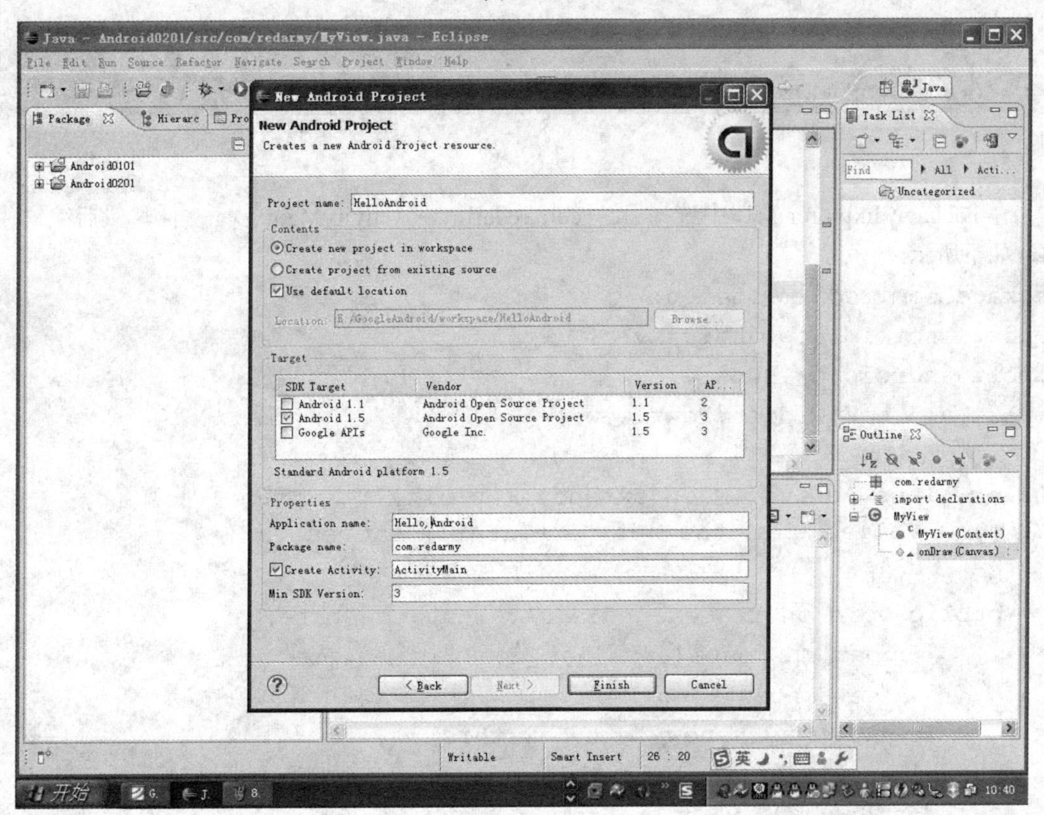
Java - Android0201/src/com/redarmy/MyView.java - Eclipse
File Edit Run Source Refactor Navigate Search Project Window Help
New Android Project
New Android Project
Creates a new Android Project resource.
Project name: HelloAndroid
Contents
Create new project in workspace
Create project from existing source
Use default location
Location: E:/GoogleAndroid/workspace/HelloAndroid
Browse...
Target
SDK Target
Vendor
Version
AP...
Android 1.1
Android Open Source Project
1.1
2
Android 1.5
Android Open Source Project
1.5
3
Google APIs
Google Inc.
1.5
3
Standard Android platform 1.5
Properties
Application name: Hello,Android
Package name: com.redarmy
Create Activity: ActivityMain
Min SDK Version: 3
< Back
Next >
Finish
Cancel
Task List
Outline
com.redarmy
import declarations
MyView
MyView(Context)
onDraw(Canvas)
Writable
Smart Insert
26 : 20
开始
10:40

图　2-35

（3）编写 Android 主程序，如图 2-36 所示。

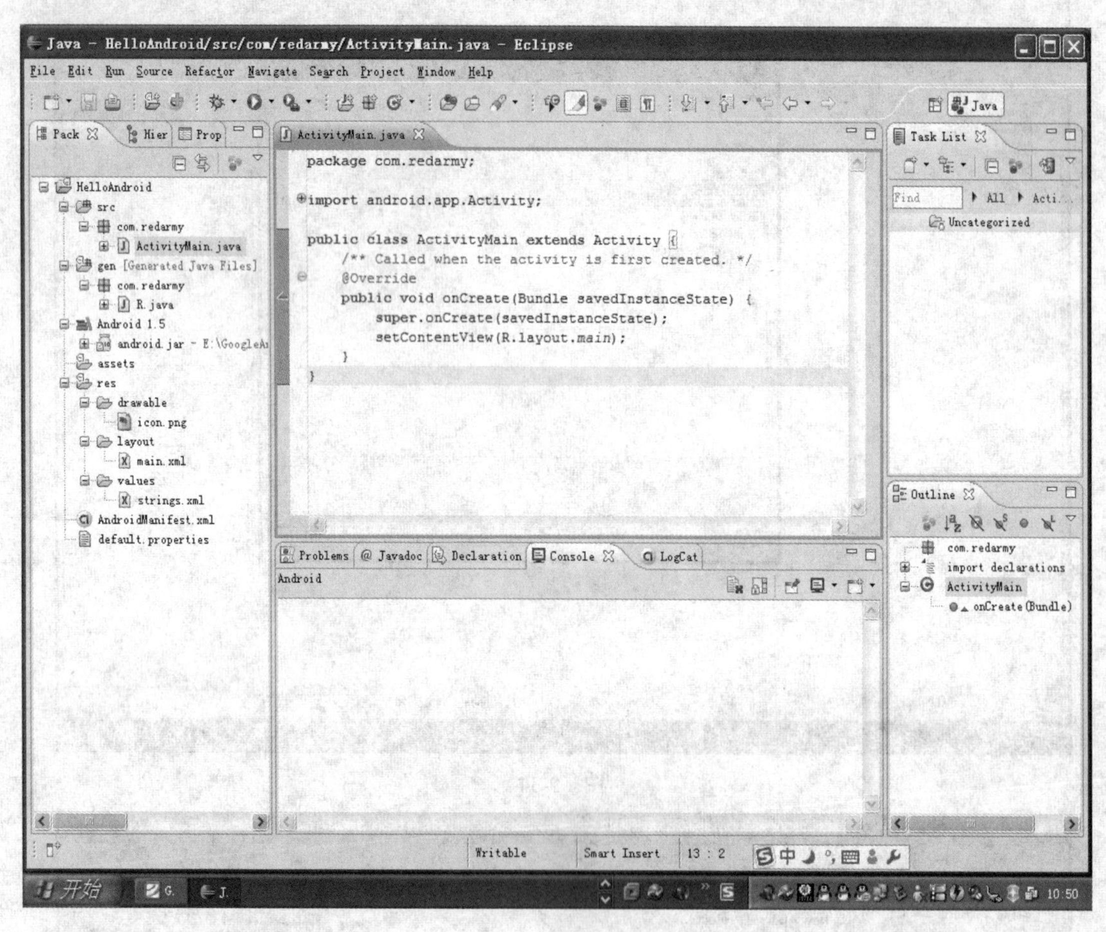

图 2-36

在 Package Explorer 窗口中选择 src→com.redarmy→ActivityMain.java 文件，然后编辑代码，如下所示：

```
package com.redarmy.hello;
import android.app.Activity;
import android.os.Bundle;
import android.widget.TextView;

public class ActivityMain extends Activity {
      /** Called when the activity is first created. */
    @Override
    public void onCreate(Bundle saveInstanceState) {
          super.onCreate(saveInstanceState);
          //setContentView(R.layout.main);
        TextView tv = new TextView(this);
          tv.setText("helloWorld");
        setContentView(tv);
```

```
    }
}
```

在 Run As 窗口中选择 Android Application，将弹出如图 2-37 所示的模拟器窗口。

单击模拟器中的 MENU 键解锁，HelloAndroid 项目的程序即显示出来，如图 2-38 所示。

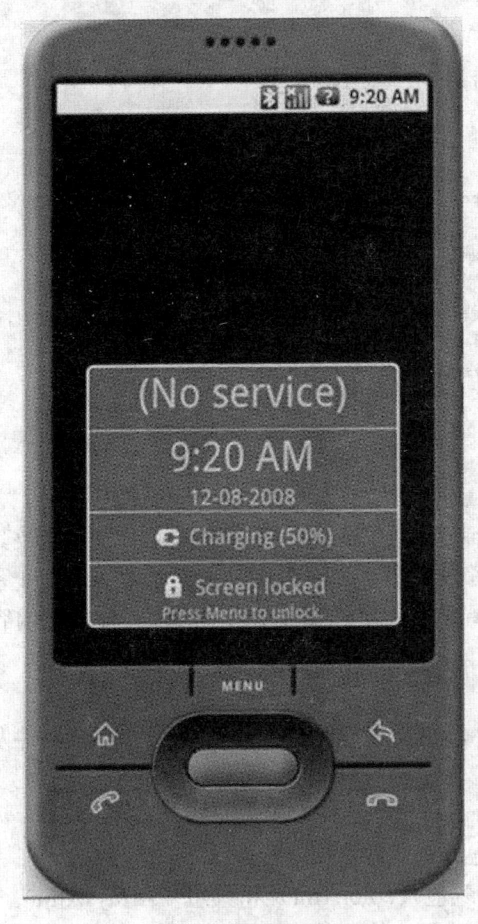

图　2-37

图　2-38

2.3.2　Android 项目调试

Android 提供的配套工具的功能也很强大，利用 Eclipse 和 Android 基于 Eclipse 的插件，可以在 Eclipse 中对 Android 程序进行断点调试。

1．设置断点

和对普通的 Java 应用设置断点一样，双击代码左边的区域可以设置断点。

2．Debug 项目

Debug Android 项目的操作和 Debug 普通的 Java 项目类似，调试项目选择 Android Application 即可。

3．断点调试

可以进行单步调试，操作步骤和调试普通 Java 程序类似。

2.3.3　Android 工程目录

Android 工程目录如图 2-39 所示，详述如下。

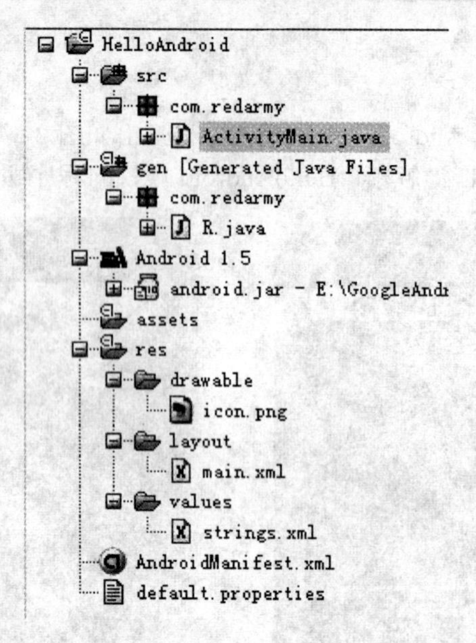

图 2-39

（1）源文件都在 src/目录中，包括活动 Java 文件和所有其他的 Java 应用程序文件。

（2）gen/包名/R.java 文件。这个文件是 Eclipse 自动生成的，应用开发者不需要修改其内容。由 Eclipse 自动处理。R 文件对于应用开发者来说基本上没有用处，但是对 Android 系统非常有用。在此文件中，Android 对资源进行全局索引。下面将要介绍的 res 文件夹的内容发生任何变化，R.java 都会重新编译，同步更新。

（3）assets/目录中主要放置多媒体文件。

（4）res/为应用程序资源，如 drawable 文件、布局文件、字符串值等。当资源文件发生变化时，R 文件的内容随之改变。

① drawable：主要放置应用到的图片资源；

② layout：主要放置用到的布局文件，都是 xml 文件；

③ values：主要放置字符串（String.xml）、颜色（color.xml）和数组（Arrays.xml）。

（5）androidMainfest.xml 文件相当于应用的配置文件。其中必须声明应用的名称，应用所用到的 Activity、Service 及 receiver 等。

第3章 Android开发准备

3.1 Android应用程序的组成

一般来说，Android程序由下列4个部分组成：

① Activity；

② Broadcast Intent Receiver；

③ Service；

④ Content Provider。

并不是所有的程序都需要这4个部分。例如，第2章介绍的程序只涉及Activity，其他3项没有涉及。这4个组件都在Android项目的AndroidManifest.xml文件里声明，并且为每个组件的功能和需求进行必要的描述。

3.1.1 Activity

一个Activity代表了可以和用户进行交互的可视化界面，通常代表手机屏幕中的一屏。如果把手机比作一个浏览器，Activity就相当于一个网页。Activity通过布局管理添加各种View组件。在Activity中可以添加View，并且对这些View做事件处理，通过setContentView(int)方法将视图呈现出来。一个应用程序由一个或者多个Activity组成，一个Android项目由多个Activity组成，并且Activity之间可以切换。

根据是否与其他Activity交互，将Activity分为两种类型：独立的和相互依赖的。

1．独立的Activity

独立的Activity不需要从其他地方取得数据，只是单纯地从一个屏幕跳到下一个屏幕，不涉及数据交换。从一个独立的Activity调用另一个独立的Activity时，只需要填写Intent的内容和动作，然后使用startActivity()函数调用，即可唤起独立的Activity。

2．相互依赖的Activity

相互依赖的Activity需要与其他Activity交换数据，又分为单向的和双向的。从一个屏幕跳转到下一个屏幕时，携带数据供下一个屏幕使用，这样的Activity是单向依赖的；在两个屏幕之间切换，屏幕上的数据因另一个屏幕的操作而改变，这样的Activity是双向依赖的。与独立的Activity相比，相互依赖的Activity变化更加复杂。

另外，每一个Activity都有生命周期（LifeCycle）。Android应用程序的生命周期由Android框架进行管理，不由应用程序直接控制。讲解Android的生命周期之前，先介绍Android的状态。Android的虚拟机（VM）使用堆叠（Stack based）管理，主要有4种状态：Active（活动）、Paused（暂停）、Stopped（停止）和Dead（已回收或未启动）。

（1）Active（活动）状态：是用户启动应用程序或Activity后，Activity运行时的状态。在Android平台上，在同一个时刻只有一个Activity处于活动（Active）或运行（Running）状态，其他Activity处于未启动（Dead）、停止（Stopped）或暂停（Pause）状态。

（2）Paused（暂停）状态：是目前运行的屏幕画面暂时暗下来，退到背景画面的状态。

当使用 Toast、AlertDialog 或电话呼入时，都会让原本运行的 Activity 退到背景画面。新出现的 Toast、AlertDialog 等界面组件将覆盖原来的 Activity 画面。Activity 处在 Paused 状态时，用户无法与原 Activity 互动。

（3）Stopped（停止）状态：是有其他 Activity 正在执行，原来运行的 Activity 已经离开屏幕，不再动作的状态。长按 Home 按钮，可以调出所有处于 Stopped 状态的应用程序列表。Stopped 状态的 Activity，可以通过 Notification 唤醒。

（4）Dead（已回收或未启动）状态：是 Activity 尚未启动，就被手动终止，或被系统回收的状态。要手动终止 Activity，在程序中调用 finish 函数。

Android 的生命周期图如图 3-1 所示。

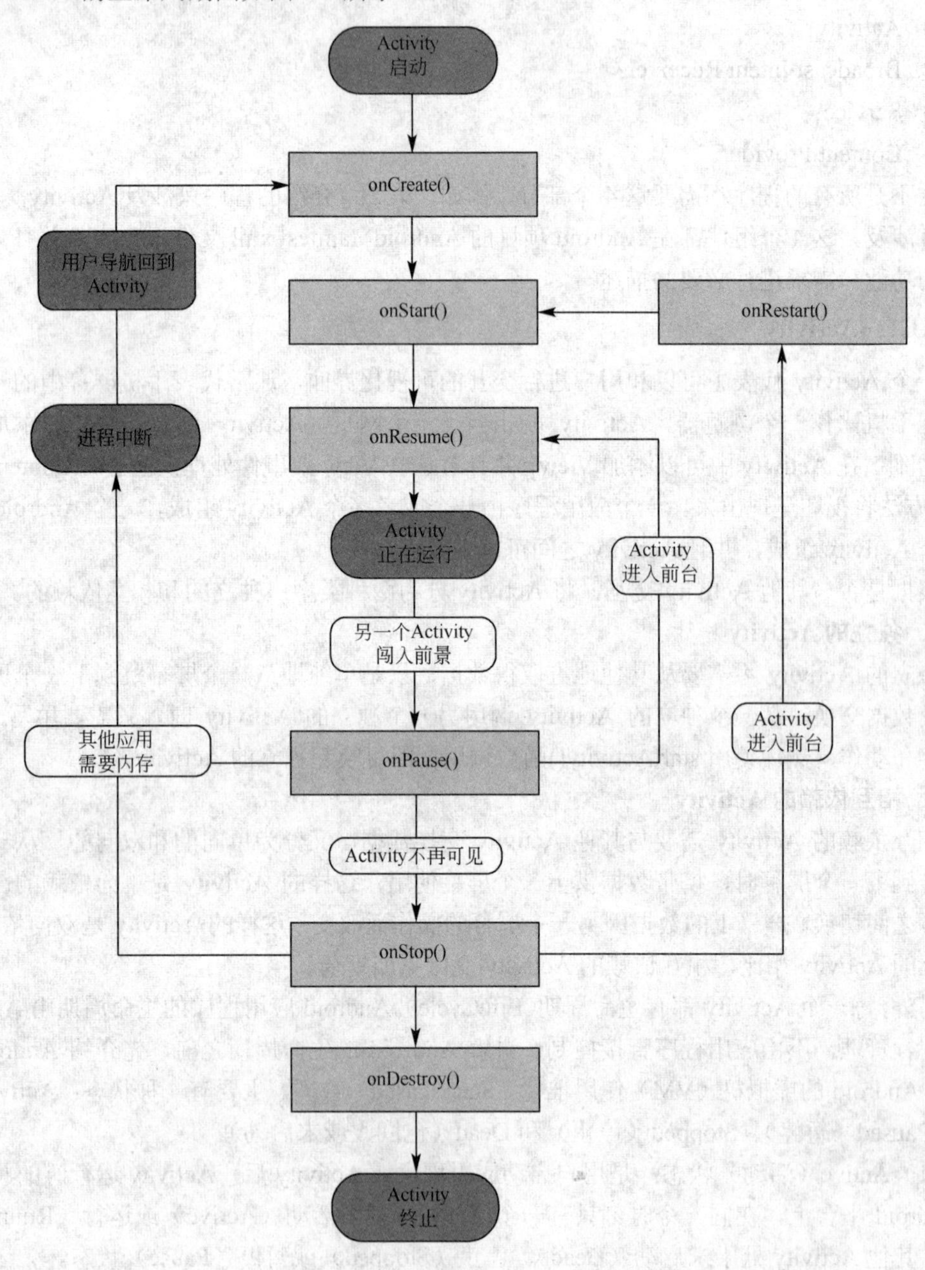

图 3-1

上述 4 种状态归纳成 3 类：资源分配（Create/Destroy）、可见与不可见（Start/ReStart/Stop）以及用户能否直接访问屏幕（Resume/Pause）。

（1）资源分配（Create/Destroy）：完整的 Activity 生命周期由 Create 状态开始，由 Destroy 状态结束。创建（Create）时分配资源，销毁（Destroy）时释放资源。

（2）可见与不可见（Start/ReStart/Stop）：当 Activity 运行到 Start 状态时，可以在屏幕上看到这个 Activity；相反地，当 Activity 运行到 Stop 状态时，Activity 从屏幕上消失。当 Activity 尚未被销毁（Destroy），而再次被调用时，进入 ReStart 状态后，再进入正常的 Start 状态。例如，从当前的 Activity 返回前一个 Activity 时，比直接打开新的 Activity 多进入一个 ReStart 状态。

（3）用户能否直接访问屏幕（Resume/Pause）：当有 Toast、AlertDialog、短信、电话等消息闯入时，原来的 Activity 进入 Pause 状态，暂时放弃直接访问屏幕的能力，被中断到背景去，将前景交给优先级高的事件。待优先级高的事件处理完毕，Activity 进入 Resume 状态，又可以直接访问屏幕。

3.1.2　Broadcast Intent Receiver

Broadcast Intent Receiver 负责对外部事件做出响应，它不生成 UI，是不可见的。Broadcast Intent Receiver 通过 NotificationManager 通知用户事件的发生，既可以在 AndroidManifest.xml 中注册，也可以在程序运行时的代码中使用 Context.registerReceivers 注册。只要注册了，当事件来临的时候，系统在需要的时候启动程序。各种应用程序通过 Context.sendBroadcast()将自己的 intent broadcasts 广播出去；其他程序通过自身的 Broadcast Intent Receiver 截获信息进行事件处理。

3.1.3　Service

Service 是一种程序，它可以运行很长时间，但是没有用户界面。Service 运行在后台，负责处理用户看不到，并且持续一段时间的事件，比如下载数据、播放音乐等。所有用户实现的 Services 必须继承系统的 Service 类，并且在配置文件中注册。

3.1.4　Content Provider

与其他操作系统不同，数据在 Android 系统中是私有的，包括文件数据和数据库数据，以及其他类型的数据。Content Provider 提供了多个程序间数据交互的机制，并提供标准的 API 对数据进行操作，简述如下。

（1）insert(Uri,ContentValues)：将一组数据插入到指定的地方。

（2）delete(Uri,String,String[])：删除数据

（3）update(Uri,ContentValues,String,String[])：更新数据。

（4）query(Uri,String[],String,String[],String)：通过关键字查询数据。

后面将深入讨论。

3.2　Android 的事件处理

3.2.1　事件监听简介

用户界面除了显示给用户观看，还需要响应用户的各种操作事件。Android 平台使用回调机制来处理用户界面事件，每个 View 都有自己的处理事件的回调方法。如果没有被 Activity 的任何一个 View 处理，Android 将调用 Activity 的事件处理回调方法进行处理。Android 平

台上一些常用事件处理的回调方法简述如下。

（1）public Boolean onKeyDown（int keyCode,KeyEvent event）：处理手机按键被按下事件的回调方法。

（2）public Boolean onKeyUp（int keyCode,KeyEvent event）：处理手机按键按下后抬起事件的回调方法。

（3）public Boolean onTrackballEvent（MotionEvent event)：处理轨迹球事件的回调方法。

（4）public Boolean onTouchEvent（MotionEvent event）：处理触摸事件的回调方法。

（5）protected void onFocusChanged（Boolean gainFocus,int direction,Rect previous）：处理焦点改变事件的回调方法。

注意：回顾 Java 中的委托模型，深刻理解 Android 的事件处理。这里简单说明委托模型的优点：

（1）好比订阅报纸，只为订阅的用户送报纸。

（2）事件的“当事人”不处理事件，避免了大量继承的烦冗。

（3）一次事件处理涉及三个对象：当事人对象、事件对象和受托人对象。

3.2.2 常用的事件监听

Android 为每个 View 提供了一种事件监听接口。每个接口需要实现一个回调方法，然后使用 View 的 setXXXXListener 方法设置事件监听接口。使用这种监听接口的优点是：当不同的 View 触发相同类型的事件时，可以调用同一种回调方法，并且不必为处理事件而重新定义控件类。常用的事件监听为：View.OnClickListener、View.OnLongClickListerner、View.OnFocusChangListerner、View.OnKeyListener、View.OnTouchListener 和 OnCreateContextMenuListener，分述如下。

1. View.OnClickListener

只有一个方法 onClick(View v)。当前 View 被单击，或者当前 View 在获得焦点的时候按下轨迹球即触发。

2. View.OnLongClickListener

只有一个方法 onClick(View v)。当前 View 被单击，或者当前 View 在获得焦点的时候，按下轨迹球即触发（超过 1s），相当于鼠标按键的功能。

3. View.OnFocusChangeListener

里边只有一个方法 onFocusChange()。当前 View 的焦点变化时，方法被调用。

4. View.OnKeyListener

只有一个方法 onKey()。当前组件获得焦点，或者用户按下键时触发。

5. View.OnTouchListener

只有一个方法 onTouch()。当触摸事件传递给当前组件时，注册在当前组件内部的 OnTouchListener 被执行。

常用代码如下所示：

```
public boolean onTouch(View v MotionEvent event){
    switch(event.getAction()){
        case MotionEvent.ACTION_DOWN:
        case MotionEvent.ACTION_MOVE:
           case MotionEvent.ACTION_UP:
```

```
        break;
    }
}
```

6. OnCreateContextMenuListener

上下文菜单显示事件监听接口。当 View 使用 showContextMenu()时，触发上下文菜单显示事件。这是定义和注册上下文菜单的另一种方式。

回调方法为：

public void onContextMenu(ContextMenu menu,View v,ContextMenuInfo info)

3.3　Intent 的简单应用

3.3.1　Intent 概述

一个 Intent 就是一次对将要执行的操作的抽象描述。Intent 的表现作用有 3 种，这里主要介绍最基本、最常用的 Intent 的作用：通过 Intent 实现多个 Activity 之间的跳转。

Intent 中有两个最要的部分：Intent 的动作（Action）和动作对应的数据（Data）。典型的动作有 MAIN（Activity 的门户）、VIEW、PICK、EDIT 等。动作对应的数据以 URI 的形式类表示。

3.3.2　Intent 实现多个 Activity 直接跳转的步骤

（1）在 Android 应用程序中创建新 Activity 和与之对应的 layout.xml 文件。

（2）在 AndroidManifest.xml 文件的<application></application>中添加：

<activity android:name="新的 Activity 名称"></activity>;

（3）在原有的 Activity 类中添加 startActivity()函数。

（4）创建新的 Intent 实体：

Intent intent = new Intent();

（5）指定来源的 Activity 所在的 class 与要跳转到的 Activity 所在的 class：

intent.setClass(ActivityMain.this, ActivityGame.class);

（6）将定义好的 Intent 添加到函数中，该函数将 Intent 传入 Android 框架。Android 框架根据各应用程序在系统中注册的信息，交给合适的 Activity 处理：

startActivity(intent);

3.4　Android 应用程序的线程模型

每一个进程（Process）有一个主线程（Main Thread）。主线程必须时时刻刻（标准为 5s 内）关注 UI 事件，以便快速响应，因而会利用子线程执行费时的工作。在 Android 平台，主线程与其子线程的分工明确。

（1）子线程负责执行费时的工作。

（2）主线程负责 UI 的操作或事件；子线程不可以插手有关 UI 的事件。

VM 是多线程的执行环境。每一个线程在呼叫 JNI 函数[如 JNI_OnLoad()函数]时，传递进来的 JNIEnv 参数所含的线程 ID 都不同。为了配合这种多线程环境，*.SO 组件开发者在撰写本地函数时，可借由 JNIEnv 参数所含的线程 ID 之不同而避免资料冲突，确保所写的本地函数能在 Android 的 VM 里安全地执行。

一个进程（Process）里有一个 Thread Pool。当多个 Client（如 Activity）要求 Bind 一个 Service 时，Binder System 从 Thread Pool 里启动一个线程来执行该 Service。而且执行 Client 的线程与执行 Service 的线程同步。

第 4 章　Android 基本组件

4.1　UI 的基本元素

本章介绍 Android 应用开发中最基础、最重要的部分——Android 基本组件。

众所周知，一个 Activity 应用是由一个或者多个 Activity 组成的。Activity 就是 UI 的容器，它本身不在用户界面显示出来。在编程之前，应当了解 UI 基本元素及其相互之间的继承结构。

4.1.1　视图组件（View）

在 Android 中，View 是最基本的 UI 类，基本上所有的高级 UI 组件都是由继承 View 类实现的。例如，TextView（文本框）、Button（按钮）、List（列表）、EditText（编辑框）、RadioButton（多选按钮）、Checkbox（选择框）等都是 View 的子类。

一个视图（View）在屏幕上占据一块矩形区域，它负责渲染这块矩形区域（如将该区域变成蓝色或其他颜色），也可以处理这个块矩形区域发生的事件（若用户单击了该区域），并且可以设置该区域是否可见，是否可以获取焦点等。

4.1.2　视图容器组件（Viewgroup）

Viewgroup 对象是 Android.view.Viewgroup 的实例。正如其名字所示，Viewgroup 的作用就是 View 的容器，它负责对添加进 Viewgroup 的 View 进行布局。

当然，Viewgroup 可以添加到另一个 Viewgroup 里，因为 Viewgroup 也继承自 View.Viewgroup 类，它是一个抽象的类，也是其他容器类的基类。下面将介绍它的一些实现类。

4.1.3　布局组件（Layout）

Viewgroups 的实现类比较多，这里只简单介绍最常用的两个：LinearLayout 和 RelativeLayout，如图 4-1 所示。LinearLayout 可以进行水平布局或者竖直布局。如果将 LinearLayout 的布局方向设置为 vertical，表示是竖直布局；如果设置为 horizontal，表示是水平布局，即从左到右依次布局。

RelativeLayout 负责相对布局。在 CSS 编程中经常用到相对布局。例如，设置 A 显示在 B 的左侧，那么 B 的显示坐标不是固定的，而是相对于 A 的位置。其他布局组件将在后面逐步介绍。

4.1.4　布局参数（LayoutParams）

将 View 加入 Viewgroup，例如加入到一个 RelativeLayout 里，它在其中如何显示？如果 RelativeLayout 的大小是 50×50，这个 View 显示在 RelativeLayout 的上边、下边，还是右边？在将每一个 View 加入 RelativeLayout 时，会传递一组值（如果没有传递值，系统采用默认值）。这组值封装在 LayoutParams 类中，在显示 View 时，其容器根据传来的 LayoutParams 进行计算，确认 View 显示的大小和位置。

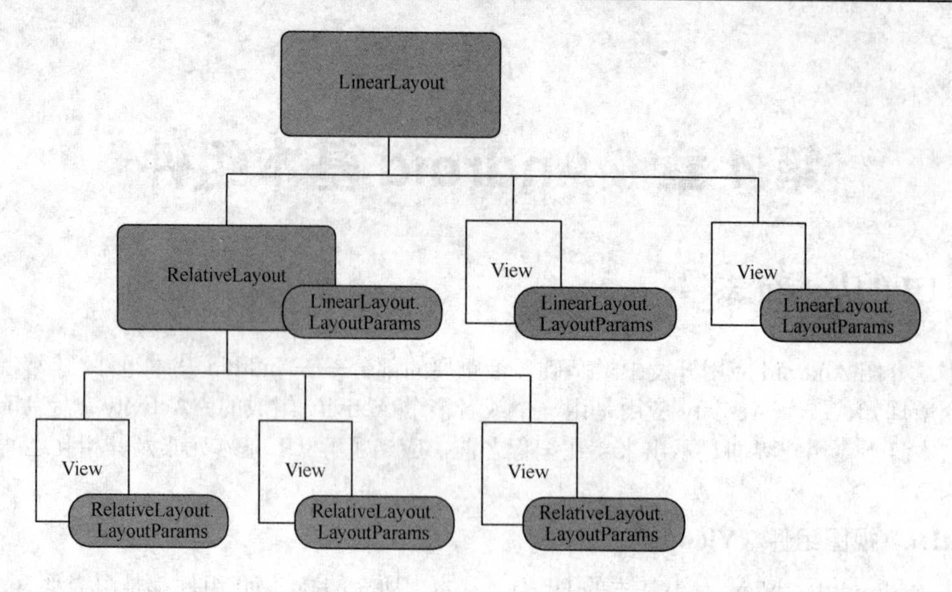

图 4-1

4.2 Android 中的 UI 布局

4.2.1 声明布局的方式

声明布局有下述两种方式：

1．在 xml 文件中声明 UI 组件

（1）Android 提供了从 xml 的节点元素，对应代码中的 UI 组件。

（2）直观简洁，可读性强。

（3）实现了 UI 界面和逻辑代码的分离。

2．在代码中构造组件

（1）在 Java 代码中构造组件。

（2）抽象模糊，可读性比较差。

（3）耦合性强，数据表现和逻辑错杂，很难修改。

4.2.2 布局属性

所有的 View 和 ViewGroup 都支持 XML 的属性，XML 的属性是可以继承的。

View 和 ViewGroup 都支持下面两个属性。

1．ID 属性

（1）android:id="@+id/my_button"

（2）android:id="@android:id/empty"

2．Layout Parameters

（1）android:layout_height

（2）android:layout_width

表达尺寸大小有下述 3 种方式：

（1）一个确定的数字（50px）。

（2）FILL_PARENT。

（3）WRAP_CONTENT。

注意：Android 支持的描述区域大小的类型如下所述：

（1）px（pixels）：像素。

（2）dip（device independent pixels）：依赖于设备的像素。

（3）sp（scaled-----best for text size）：带比例的像素。

（4）pt（points）：点。

（5）in（inches）：英尺。

（6）mm（millimeters）：毫米。

4.2.3　Android 中的盒子模型

盒子模型是非常重要的概念，如图 4-2 所示。如果学过 DIV+CSS 的知识，比较容易理解。这里只说明 Android 中盒子模型的注意事项。

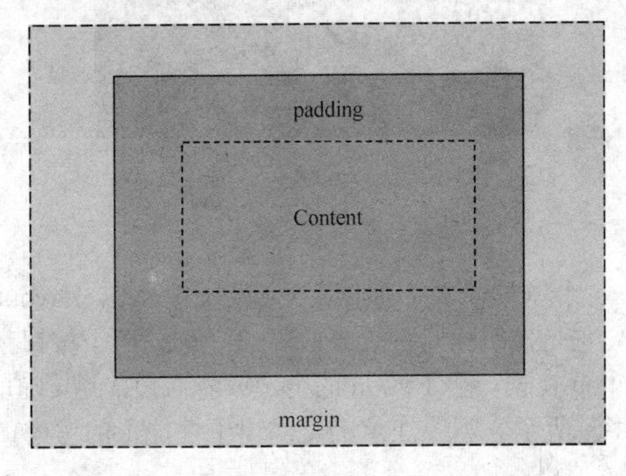

图　4-2

（1）View 支持 padding，但是不支持 margin。

（2）ViewGroup 支持 padding 和 margin。常用的方法有：

① setPadding(int,int,int,int);

② getPaddingLeft();

③ getPaddingTop()。

4.2.4　Android 中常见的布局

Android 常用的布局有 LinearLayout、FrameLayout、RelativeLayout、TableLayout 和 AbsoluteLayout，分述如下。

1．LinearLayout

LinerLayout 是 Android 中最常用的布局之一，它将自己包含的子元素按照一个方向进行布局排列，如图 4-3 所示。有以下两种方向：

（1）水平（Android:orientation="horizontal";）：子元素从左到右一个一个地水平排列。

（2）竖直（Android:orientation="vertical";）：子元素从上到下一个接着一个地竖直排列。

2．FrameLayout

FrameLayout 对象好比一块在屏幕上预定的空白区域，可以填充一些元素。

注意：所有元素都放置在 FrameLayout 区域的左上角，无法为这些元素指定确切的位置。如果一个 FrameLayout 里，有多个子元素，后边的子元素将重叠显示在前一个元素上。

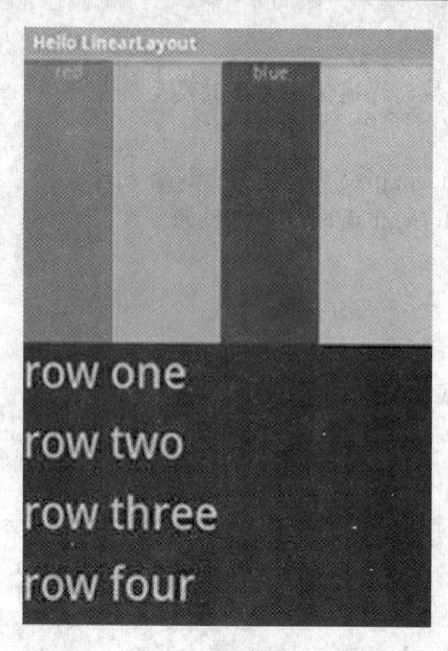

图 4-3

3．RelativeLayout

RelativeLayout 是一个相对布局类。如图 4-4 所示。RelativeLayout 是一个容器，其中元素，如 Button 按钮等的位置是按照相对位置来计算的。例如，两个按钮都在一个 Relativelayout 里，可以定义第二个 Button 在第一个 Button 的上边或者右边，但其确切的位置依赖于第一个 Button 的位置，如图 4-5 所示。需要注意的是，出于性能上的考虑，对于相对布局精确位置的计算只会执行一次，所以，如果一个可视化组件 B 依赖于 A，那么必须让 A 出现在 B 之前。

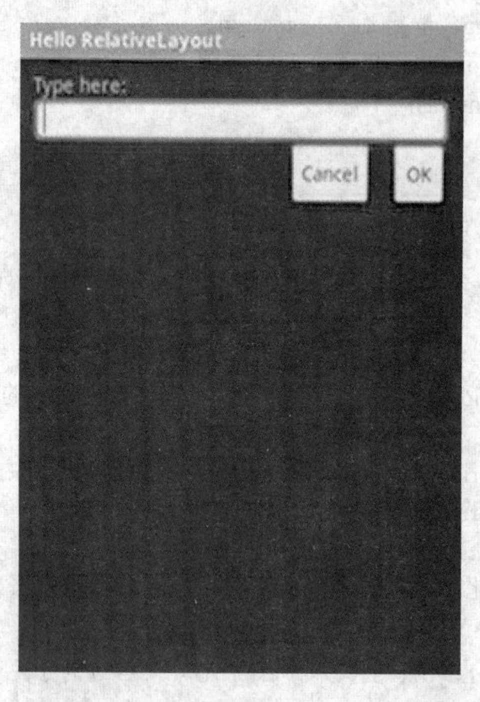

图 4-4

让子元素(通过 ID)指定它们相对于其他元素的位置或相对于父布局对象

Activity Title property

Views/Layouts/RelativeLayout/Example

Type here:

Cancel　OK

TextView
- ID: label1
- width: Fill Parent
- height: wrap content
- text: Type here:

EditText
- ID: textEntry
- width: Fill Parent
- height: wrap content
- below: "label1"

RelativeLayout
- width: Fill Parent
- height: wrap content
- background: blue
- padding: 10

Button
- ID: okButton
- width: wrap content
- height: wrap content
- below: "textEntry"
- alignParentRight: true
- marginLeft: 10
- text: "OK"

Button
- width: wrap content
- height: wrap content
- toLeft: "okButton"
- alignTop: "okButton"
- text: "Cancel"

图　4-5

4．TableLayout

TableLayout 是一种表格式布局，它把包含的元素以行和列的形式进行排列。表格的列数为每一行的最大列数。当然，表格里的单元格可以为空，如图 4-6 所示。

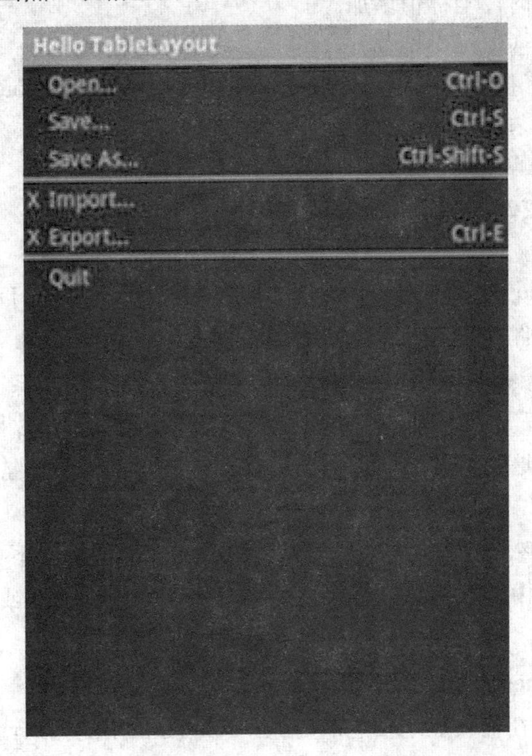

图　4-6

5．AbsoluteLayout

AbsoluteLayout 使子元素能够指明确切的（*x*,*y*）坐标并显示在屏幕上，如图 4-7 所示。

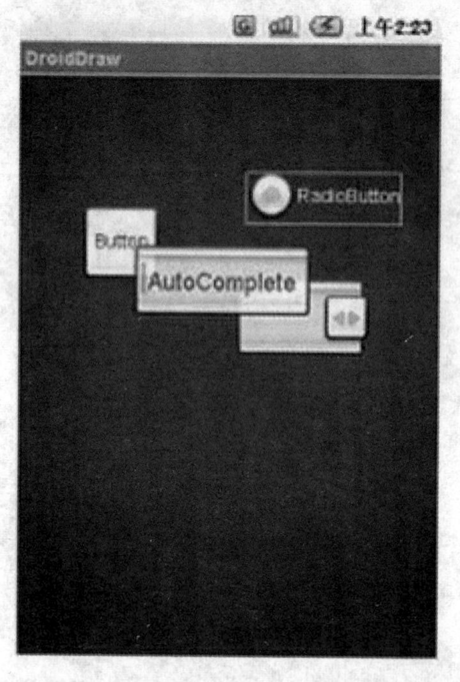

图 4-7

（1）（0,0）是左上角。

（2）当下移或右移时，坐标值增加。

（3）允许元素重叠（但是不推荐）。

注意：除非有很好的理由，一般建议不使用 AbsoluteLayout。因为它相当严格，并且在不同的设备显示中不能很好地工作。

4.3 常用的 Widget 组件

前面介绍了 Android 的组件关系，讲解了如何编写一个界面的布局。下面介绍布局中的组件。只有了解布局组件的使用方法，才能更合理地实现界面布局。

Android 中提供了一种叫做 Widget 的包，其中包含各种 UI 元素（大部分是可见的），在应用程序的屏幕中使用，例如按钮、文本框、编辑框、进度条、图片显示等。在进行 Android 编程之前，应大致了解这类组件，知道每个组件的样式、使用的场景及使用方法。

1．按钮（Button）

Button（按钮）在整个组件中的地位非常特殊，它在一般情况下扮演“终结者”的角色。例如，用户在一个界面中输入信息，很多情况下都有一个“确认”或者“取消”按钮，用户的动作一般在这里结束，然后开始另外一系列动作。

2．文本框（TextView）

TextView 一般使用在需要显示信息的时候。TextView 不能输入，只能初始设定或者在程序中修改。如果需要在程序中动态地修改这个值，应使用其 Android:id 的值，代码如下：

```
public void find_and_modify_textView(){
   TextView text_view=(TextView)findViewById(R.id.my_textView);
   CharSequence text_view_old = text_view.getText();
   text_view.setText("修改为:TextView 的值也是可以动态修改的.");
}
```

3．编辑框（EditText）

EditText 是文本编辑框，提供一个输入框以便用户输入。它通过编程事件处理，用 getText() 获取用户输入值，代码如下所示：

```
public void onClick(View v) {
  EditText edit_text = (EditText)findViewById(R.id.my_editText);
      CharSequence editTextValue = edit_text.getText();
      setTitle("EditText 的值: "+editTextValue);
}
```

4．多项选择（CheckBox）

除了 EditText 组件外，CheckBox 也是一个使用频繁的组件。CheckBox 一般用来提供给用户输入信息，可以一次选择多个选项。当在手机屏幕上不方便执行输入操作时，使用选择组件来帮助用户单击输入选项。CheckBox 提供了 isChecked()方法来识别选中或未选中状态。返回值如果为真，表示选中，否则表示未选中。

5．单选择（RadioGroup）

RadioGroup 提供了一种多选一的选择模式，也是非常有用的组件。可以使用 RadioGroup 提供的 clearCheck()来清除选项。

6．下拉列表（Spinner）

Spinner 提供一种下拉列表选择的输入方式，在手机应用界面非常常见，可以节省有限的屏幕空间占用。一个 Spinner 下拉组件在界面中显示，但是没有任何数据。给 Spinner 组件装填数据常用的两种方式如下所述。

（1）在编程的时候载入列表数据。

首先，定义以下需要装载的数据：

```
private static final String[] mCountries = { "China" ,"Russia",
                    "Germany","Ukraine", "Belarus", "USA" };
```

其次，在 onCreate 方法中调用装载数据的方法：

```
private void find_and_modify_view() {
spinner_c = (Spinner) findViewById(R.id.spinner_1);
      allcountries = new ArrayList<String>();
      for (int i = 0; i < mCountries.length; i++) {
         allcountries.add(mCountries[i]);
      }
      aspnCountries = new ArrayAdapter<String>(this,
            android.R.layout.simple_spinner_item, allcountries);
      aspnCountries.setDropDownViewResource(android.R.layout.
simple_spinner_dropdown_item);
      spinner_c.setAdapter(aspnCountries);
   }
```

上述代码将自定义 mCountries 数据装载到 Spinner 组件中。

（2）在 XML 文件中预先定义数据

除了上述指定 Spinner 内容的方法之外，还可以在 XML 文件中预先定义数据。为了说明该用法，在 spinner.xml 模板中添加一个 Spinner 组件，代码如下所示：

```
<TextView android:layout_width="fill_parent"
          android:layout_height="wrap_content"
          android:text="Spinner_2 From arrays xml file"/>
<Spinner  android:id="@+id/spinner_2"
          android:layout_width="fill_parent"
          android:layout_height="wrap_content"
          android:drawSelectorOnTop="false"/>
```

然后，在 SpinnerAcitivity.java 中初始化其值，代码如下所示：

```
spinner_2 = (Spinner) findViewById(R.id.spinner_2);
ArrayAdapter<CharSequence> adapter =
   ArrayAdapter.createFromResource(this, R.array.countries,
                               android.R.layout.simple_spinner_item);
   dapter.setDropDownViewResource(android.R.layout.simple_spinner_
dropdown_item);
   spinner_2.setAdapter(adapter);
```

上述代码将 R.array.countries 对应的值载入 Spinner_2。R.array.countries 对应的值在 XML 文件中预先定义，方法是在 res/values/目录下新建一个名为 array.xml 的文件，内容如下：

```
<?xml version="1.0" encoding="utf-8"?>
<resources>
   <!-- Used in Spinner/spinner_2.java -->
   <string-array name="countries">
      <item>China2</item>
      <item>Russia2</item>
      <item>Germany2</item>
      <item>Ukraine2</item>
      <item>Belarus2</item>
      <item>USA2</item>
   </string-array>
</resources>
```

7. 自动完成文本（AutoCompleteTextView）

除了上述 Spinner 组件外，Android 还提供了一种节省界面空间的辅助输入方式 AutoCompleteTextView，但是需要将数据绑定到该组件，方法如下所述。

首先，定义要绑定的数据：

```
static final String[] COUNTRIES = new String[] {
   "China" ,"Russia", "Germany",
   "Ukraine", "Belarus", "USA" ,"China1" ,"China12", "Germany",
   "Russia2", "Belarus", "USA"
   };
```

然后，在 onCreate 中绑定，代码如下所示：

```
ArrayAdapter<String> adapter = new ArrayAdapter<String>(this
                 ,android.R.layout.simple_dropdown_item_1line
                 , COUNTRIES);
AutoCompleteTextView textView = (AutoCompleteTextView)
                 findViewById(R.id.auto_complete);
textView.setAdapter(adapter);
//根据定义的COUNTRIES创建ArrayAdapter,并将其绑定到AutoCompleteTextView组件。
```

AutoCompleteTextView 的自动完成功能提供了非常好的用户体验。若想进一步学习，请仔细阅读 Android SDK 文档中的相关内容。

8．日期选择器（DatePicker）

DatePicker 是一个日期选择组件，用于提供快速选择日期的方式。DatePicker 的展示方式和使用方式有很多种，这里简单介绍。

```
DatePicker dp=(DatePicker)this.findViewById(R.id.date_picker);
dp.init(2009, 5, 17, null);    //设定开始时间
```

9．时间选择器（TimePicker）

TimePicker 是时间选择组件，用于提供快速选择和调整时间的方式。TimePicker 的属性比较多，定义不同的属性，表现的外观差异比较大，代码如下所示：

```
TimePicker tp=(TimePicker)this.findViewById(R.id.time_picker);
tp.setIs24HourView(true);
```

10．滚动视图（ScrollView）

在屏幕中添加组件，如果达到一定的程度，会超出屏幕的高度范围，若再增加按钮或内容，将被挡住而看不见了。若希望在屏幕显示内容不完全的情况下，可以往下滚动，显示出被挡住的部分，则需要使用 ScrollView 组件。

ScrollVeiw 的功能主要是将屏幕显示不了的内容，通过滚动而显示出来。使用该组件比较直观，直接在布局外面增加 ScrollView 组件声明即可，代码如下所示：

```
<ScrollView
    xmlns:android="http://schemas.android.com/apk/res/android"
    android:layout_width="fill_parent"
    android:layout_height="wrap_content">
</ScrollView>
```

11．进度条（ProgressBar）

ProgressBar 是个非常有用的组件，其最直观的感觉就是进度条显示。但是在 Android 中，进度条有很多种，这里只介绍两种，代码如下所示：

```
<?xml version="1.0" encoding="utf-8"?>
<LinearLayout
   xmlns:android="http://schemas.android.com/apk/res/android"
   android:orientation="vertical"android:layout_width="fill_parent"
   android:layout_height="wrap_content">
   <TextView android:layout_width="wrap_content"
             android:layout_height="wrap_content"
```

```
            android:text="圆形进度条" />
    <ProgressBar android:id="@+id/progress_bar"
            android:layout_width="wrap_content"
            android:layout_height="wrap_content"/>
    <TextView android:layout_width="wrap_content"
            android:layout_height="wrap_content"
            android:text="水平进度条" />
    <ProgressBar android:id="@+id/progress_horizontal"
        style="?android:attr/progressBarStyleHorizontal"
            android:layout_width="200dip"
            android:layout_height="wrap_content"
            android:max="100"                    //进度最大值 100
            android:progress="50"                //第一个进度到 50
            android:secondaryProgress="75" />    //第二个进度到 75
</LinearLayout>
```

定义的进度条可以通过代码实时更新。例如在资源下载等场景，进度条提示下载进度。

12．拖动条（SeekBar）

SeekBar 组件和水平 ProgressBar 组件的功能相似，不同点在于 SeekBar 可以拖动。其代码如下所示：

```
<?xml version="1.0" encoding="utf-8"?>
<LinearLayout
    xmlns:android="http://schemas.android.com/apk/res/android"
    android:orientation="vertical"android:layout_width="fill_parent"
    android:layout_height="wrap_content">
    <TextView   android:layout_width="wrap_content"
        android:layout_height="wrap_content"
        android:text="SeekBar" />
    <SeekBar android:id="@+id/seek"
        android:layout_width="fill_parent"
        android:layout_height="wrap_content"
        android:max="100"
        android:thumb="@drawable/seeker"
        android:progress="50"/>
</LinearLayout>
```

13．评分组件（RatingBar）

在让用户参与评分的应用中，使用 RatingBar 组件非常方便，既便于用户输入，又很直观。

参考代码如下所示：

```
   <?xml version="1.0" encoding="utf-8"?>
<LinearLayout xmlns:android="http://schemas.android.com/apk/res/android"
   android:orientation="vertical" android:layout_width="fill_parent"
   android:layout_height="wrap_content">
```

```
<TextView
      android:layout_width="wrap_content"
      android:layout_height="wrap_content"
      android:text="RatingBar" />
 <RatingBar android:id="@+id/rating_bar"
 android:layout_width="wrap_content"
 android:layout_height="wrap_content"
 ratingBarStyleSmall="true" />
</LinearLayout>
```

14．图片视图（ImageView）

ImageView 主要用于展示图片，适用于很多场合。其主要属性 Android:src 为一张图片，位于项目根目录下 src 的 drawable，支持 PNG、JPG、GIF 等图片格式。

参考代码如下所示：

```
 <?xml version="1.0" encoding="utf-8"?>
<LinearLayout xmlns:android="http://schemas.android.com/apk/res/android"
   android:orientation="vertical" android:layout_width="fill_parent"
   android:layout_height="wrap_content">
<TextView
      android:layout_width="wrap_content"
      android:layout_height="wrap_content"
      android:text="图片展示:" />
<ImageView
 android:id="@+id/imagebutton"
 android:src="@drawable/eoe"
 android:layout_width="wrap_content"
 android:layout_height="wrap_content"/>
</LinearLayout>
```

15．图片按钮（ImageButton）

ImageButton 组件提供图片按钮，其操作与 ImageView 类似。参考代码如下所示：

```
 <?xml version="1.0" encoding="utf-8"?>
<LinearLayout
xmlns:android="http://schemas.android.com/apk/res/android"
   android:orientation="vertical" android:layout_width="fill_parent"
   android:layout_height="wrap_content">
<TextView
      android:layout_width="wrap_content"
      android:layout_height="wrap_content"
      android:text="图片按钮:" />
<ImageButton id="@+id/imagebutton"
 android:src="@drawable/play"
 android:layout_width="wrap_content"
```

```
  android:layout_height="wrap_content"/>
</LinearLayout>
```

4.4 菜单（Menu）

除了直接显示在 Activity 主界面中的用户控件外，Android 还有比较常见的用户界面元素，分别是菜单和对话框。Android 手机用一个按键 Menu 专门来显示菜单，所以，若应用程序设置了菜单，便可以通过该按键来操作应用程序的菜单选项。下面将深入研究 Android 中的 Menu。

4.4.1 菜单（Menu）简介

想要让 Android 程序有完善的用户体验，除了设计人性化的用户界面以外，添加一些菜单也是必要的。菜单提供额外的选项，使用户有更多的功能选择；同时，用户可以通过菜单对当前程序的功能进行更多的设置。Android 平台提供的菜单大体分为 3 类：选项菜单（Option Menu）、上下文菜单（Context Menu）和子菜单（Submenu）。下面分别介绍这 3 种菜单的概念、创建及使用方法。

1．选项菜单

如果在 Activity 主界面按手机上的 Menu 键，屏幕底部将弹出相应的带图标的选项菜单，如图 4-8 所示。这种带图标的菜单最多只能显示 6 个菜单项，若设置了 6 个以上的菜单项，弹出的带图标的选项菜单只显示前 5 个，右下角的第 6 个是“更多（More）”菜单项，如图 4-8 右上角所示。单击“更多（More）”菜单项，将显示一个浮于主界面之上的扩展选项菜单，如图 4-8 左下角所示。扩展选项菜单不支持显示图标，但可显示单选框和复选框，如图 4-8 右下角所示。

图 4-8

（1）创建选项菜单。如何定义 Activity 的选项菜单呢？第一次调用选项菜单时，Activity 调用方法 onCreateOptionMenu()，所以只需重写（Override） onCreateOptionMenu()方法，并初始化选项菜单即可。这里举一个简单的选项菜单初始化的例子，如图 4-9 所示，代码如下所示：

图　4-9

```
    private final int MENU_SAVE = Menu.FIRST;
    private final int MENU_DELETE = Menu.FIRST + 1;
   @Override
   public boolean onCreateOptionsMenu(Menu menu) {
       super.onCreateOptionsMenu(menu);
       menu.add(0, MENU_SAVE, 0, "保存").setIcon(android.R.drawable.
ic_menu_save);
      menu.add(0, MENU_DELETE, 0, "删除").setIcon(android.R.drawable.
ic_menu_delete);
       return true;
   }
```

可以看到，在 onCreateOptionMenu 方法中调用了传入的菜单（Menu）的 add 方法来添加菜单项。例中的 add 方法在添加菜单项的同时，指定了菜单项所属菜单组的 id、自己的 id、排列的顺序和显示的标题。

需要注意的是，菜单项的 id 必须是唯一的，因为在处理选项菜单的事件时，是通过菜单项的 id 来区分哪一个菜单项被单击了。通常用从 Menu.FIRST 递增的数字为每个菜单项分配唯一的 id。

add 方法返回一个菜单项的实例（MenuItem），可以通过该实例进一步设置菜单项。上例中，直接通过它调用 setIcon 方法来设置显示的图标。表 4-1 列出了 MenuItem 常用的可选设置。

表　4-1

可选设置	设置方法

单选框与复选框	单选框：直接用 MenuItem 调用 setCheckable(true)
	复选框：需要与菜单组 id 配合使用，示例如下：
	final int R_GP = 0;
	final int R_1 = Menu.FIRST;
	final int R_2 = Menu.FIRST + 1;
	menu.add(R_GP, R_1, 0, “Radiobutton 1”);
	menu.add(R_GP, R_2, 0, “Radiobutton 2”);
	menu.setGroupCheckable(R_GP, true, true);

快捷键	设置快捷键后，在选项菜单弹出的情况下，通过按快捷键直接选择需要的菜单项。采用下面 3 个方法都可以设定快捷键： （1）设置数字快捷键： setNumericShortcut(char numericChar) （2）设置数字和字符快捷键： setShortcut(char numeericChar, char alphaChar) （3）设置字符快捷键： setAlphabeticShortcut(char alphaChar)

续表

<table>
<tr><th>可选设置</th><th>设置方法</th></tr>
<tr><td>短标题</td><td>当标题太长，有可能显示不全时，可以用短标题代替，调用 setTitleCondensed（CharSequence title）设置</td></tr>
<tr><td>图标</td><td>设置菜单项的图标，不会在扩展菜单显示，调用 setIcon（int iconRes）或 setIcon（Drawable icon）设置</td></tr>
<tr><td>监听菜单项单击</td><td>这是可选的一种处理菜单项单击事件的方法，用匿名类实现，设置代码如下所示：
setOnMenuItemClickListener(new OnMenuItemClickListener(){
    @Overrider
    public Boolean onMenuItemClick(MenuItem item){
        //根据 id 判断哪个菜单项被单击，并做相应的处理
        return true;
    }
}</td></tr>
<tr><td>设置 Intent</td><td>一个菜单项可以直接与一个 Intent 直接相关联。当菜单项的单击事件没有被处理时，Activity 传入设置的 Intent，并调用 startActivity(Intent i)方法启动一个 Activity。调用 setIntent(Intent i)设置</td></tr>
</table>

（2）监听选项菜单。除了上述给菜单项设置监听器的方式用于处理菜单项的单击事件以外，重写 onOptionsItemSelected()方法可以实现同样的功能。下述代码用于处理初始化的菜单。

```
@Override
public boolean onOptionsItemSelected(MenuItem item) {
    switch(item.getItemId()){
    case MENU_SAVE :
        //做相应的处理
        break;
    case MENU_DELETE :
        //做相应的处理
        break;
    }
    return super.onOptionsItemSelected(item);
}
```

第一次初始化选项菜单后，如果需要动态更改，应重新实现 onPrepareOptionsMenu 方法，在每次显示选项菜单之前调用。可以在此方法里根据程序的运行情况即时地更新菜单项内容，如标题、是否可用等。动态更新选项菜单的实例代码如下所示：

```
@Override
public boolean onPrepareOptionsMenu(Menu menu) {
    super.onPrepareOptionsMenu(menu);
    //通过 id 找到 MenuItem
    MenuItem item = menu.findItem(MENU_SAVE);

    //根据需要，更改 item
    return true;
}
```

2．上下文菜单

上下文菜单悬浮于主界面之上。当注册到一个 View 对象上以后，在默认情况下，用户可以通过长按(约 2s)View 对象，呼出上下文菜单。上下文菜单的每个元素依然是菜单项，但它不支持显示图标和设置快捷键。另外，上下文菜单可设置顶部标题和顶部图标，如图 4-10 所示。

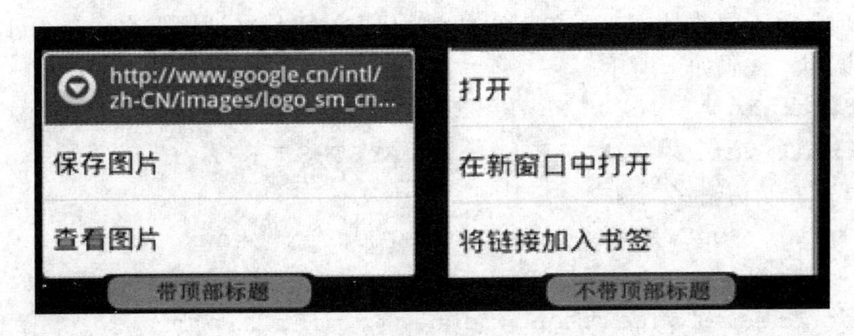

图　4-10

与选项菜单类似，初始化上下文菜单依然是通过重写 onCreateContextMenu 方法来完成，处理上下文菜单的菜单项的单击事件是通过重写 onContextItemSelected 方法来完成。如果要为 View 对象注册上下文菜单，使用 registerForContextMenu 方法，如图 4-11 所示，代码如下所示：

图　4-11

```
public class Test extends Activity {
    @Override
    public void onCreate(Bundle savedInstanceState) {
        super.onCreate(savedInstanceState);

        TextView v = new TextView(this);
        v.setText("上下文菜单的实现");
        registerForContextMenu(v);
        setContentView(v);
    }
```

首先，在程序一开始定义一个 TextView，作为呼出上下文菜单的载体，并给它注册上下文菜单，代码如下：

```
public void onCreate(Bundle savedInstanceState) {
    super.onCreate(savedInstanceState);
    TextView v = new TextView(this);
    v.setText("上下文菜单的实现");
    registerForContextMenu(v);
```

```
    setContentView(v);
}
```

然后，重写前述两个方法来初始化上下文菜单和处理单击事件。依然用 add 方法添加菜单项。与选项菜单不同的是，上下文菜单的初始化方法不只被调用一次，会在每次呼出上下文菜单时被调用，代码如下所示：

```
private final int C_MENU_NEW = Menu.FIRST + 2;      //定义 id
private final int C_MENU_OPEN = Menu.FIRST + 3;
@Override   //初始化菜单
public void onCreateContextMenu(ContextMenu menu, View v,
ContextMenuInfo menuInfo) {
    super.onCreateContextMenu(menu, v, menuInfo);
    menu.add(0,C_MENU_NEW,0,"新建");
    menu.add(0,C_MENU_OPEN,0,"打开");
}
@Override    //事件监听
public boolean onContextItemSelected(MenuItem item) {
    switch(item.getItemId()){
    case C_MENU_NEW :
        //做出相应的处理
        break;
    case C_MENU_OPEN :
        //做出相应的处理
        break;
    }
    return super.onContextItemSelected(item);
}
```

3. 子菜单

子菜单是可以被添加到其他菜单上的菜单，但是子菜单不能添加到子菜单上。通常，当有大量菜单项需要显示时，可以利用子菜单对这些菜单项进行分类，便于用户更容易找到所需的菜单项，也使菜单看起来更整洁，更有条理。在 Android 中使用子菜单十分简单，只需要把添加菜单项的 add 方法替换成 addSubMenu 方法，如图 4-12 所示，代码如下所示：

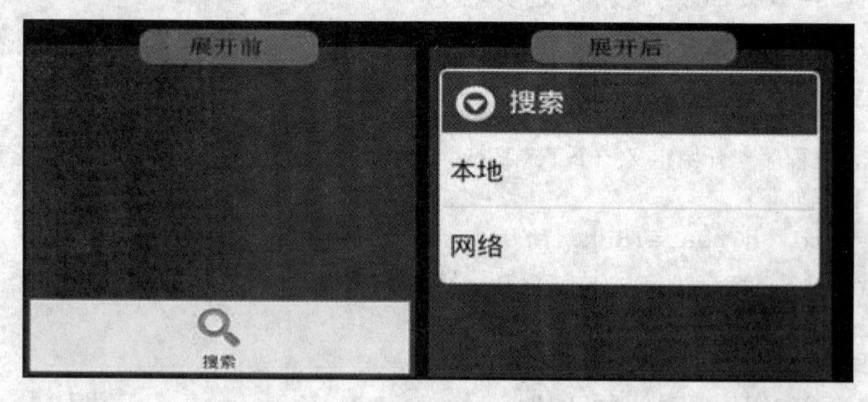

图 4-12

```
private final int MENU_LOCAL = Menu.FIRST + 4;
private final int MENU_NET = Menu.FIRST + 5;

@Override
public boolean onCreateOptionsMenu(Menu menu) {
    super.onCreateOptionsMenu(menu);

    SubMenu sub = menu.addSubMenu("搜索");
    sub.setIcon(android.R.drawable.ic_menu_search);
    sub.add(0, MENU_LOCAL, 0, "本地");
    sub.add(0, MENU_NET, 0, "网络");
    return true;
}
```

本段代码中需要说明的是：

```
SubMenu sub = menu.addSubMenu("搜索");
```

这里的 addSubMenu 方法返回一个子菜单项的实例，通过它可以给子菜单添加菜单项，代码如下所示：

```
sub.add(0,MENU_LOCAL,0,"本地");
sub.add(0,MENU_NET,0,"网络");
```

4.4.2 菜单（Menu）的创建方法

要实现菜单的功能，首先要创建菜单。上面用 oCreateOptionsMenu 创建了菜单，并且简单地介绍了几种菜单的监听方法。也可以通过 XML 的布局来实现，分述如下。

1. 通过 xml 布局来实现

第一步：首先在项目目录 res/下新建 menu 文件夹，并在其下新建/menu.xml 文件，然后，编写菜单代码如下所示：

```
<menu xmlns:android="http://schemas.android.com/apk/res/android">
    <item android:id="@+id/about" android:title="@string/about"></item>
    <item android:id="@+id/exit" android:title="@string/exit"></item>
</menu>
```

第二步：在代码中加载 xml 定义的菜单布局文件，如下所示：

```
public class ActivityMain extends Activity {
    /** Called when the activity is first created. */
    @Override
    public void onCreate(Bundle savedInstanceState) {
        super.onCreate(savedInstanceState);
        setContentView(R.layout.main);
    }

    @Override
    public boolean onCreateOptionsMenu(Menu menu) {
        MenuInflater inflater = getMenuInflater();
```

```
        // 设置menu界面为res/menu/menu.xml
        inflater.inflate(R.menu.menu, menu);
        return true;
    }
}
```

第三步：修改 values 下的 strings.xml 文件，代码如下所示：

```
<?xml version="1.0" encoding="utf-8"?>
<resources>
    <string name="exit">退出</string>
    <string name="about">关于</string>
    <string name="hello">Hello World, ActivityMain!</string>
    <string name="app_name">Menu</string>
</resources>
```

直接运行，就能看到效果。

2．通过 Menu.add 方法实现

该方法是在 OnCreateOptionsMenu 方法中通过 menu.add（0, MENU_SAVE, 0, "保存"）来创建一个菜单选项，代码如下所示：

```
public class ActivityMenu extends Activity {
    private static final int MENU_SAVE = Menu.FIRST;
    private static final int MENU_DELETE = Menu.FIRST + 1;
    @Override
    public void onCreate(Bundle savedInstanceState) {
        super.onCreate(savedInstanceState);
        setContentView(R.layout.main);
    }
    @Override
    public boolean onCreateOptionsMenu(Menu menu) {
        super.onCreateOptionsMenu(menu);
        menu.add(0, MENU_SAVE, 0, "保存")
                .setIcon(android.R.drawable.ic_menu_save);
        menu.add(0, MENU_DELETE, 0, "删除").setIcon(
                android.R.drawable.ic_menu_delete);
        return true;
    }
}
```

直接运行，就能看见效果。

在上述菜单中用到.setIcon()方法来添加选项的图标，或者使用.setAlphabeticShortcut（'英文字母'）方法来指定快捷键，也可以使用内置的菜单图标方法 android.R.drawable.ic_menu_help。这里不再赘述，读者可多加练习。

4.4.3 菜单（Menu）的事件处理

创建菜单的方式不同，处理事件的方式不尽相同，但都采用同一个事件处理方法。如果采用 XML 布局创建菜单，定义的菜单项 id 保存到 R 类文件里。单击菜单，框架捕获被单击的 MenuItem 组件，传递给相应的处理方法，代码如下所示：

```
// 处理菜单事件
   @Override
   public boolean onOptionsItemSelected(MenuItem item) {
       // 得到当前选中的 MenuItem 的 ID
       int item_id = item.getItemId();
       switch (item_id) {
       case R.id.about:
//         Toast.makeText(this, "您选中的是关于菜单", Toast.LENGTH_
SHORT).show();
           //在这里也可以跳转到其他 Activity 中
           /*新建一个 Intent 对象*/
           Intent intent = new Intent();
           /*指定一个要启动的类*/
           intent.setClass(ActivityMain.this, ActivityMenu.class);
           /*启动一个新的 Activity*/
           startActivity(intent);
           /*关闭当前的 Activity*/
           ActivityMain.this.finish();
           break;
       case R.id.exit:
           ActivityMain.this.finish();
           break;
       }
       return true;
   }
```

如果采用 Menu.add()方法处理事件，代码如下所示：

```
// 处理菜单事件
   @Override
   public boolean onOptionsItemSelected(MenuItem item) {
       // 得到当前选中的 MenuItem 的 ID
       int item_id = item.getItemId();
       switch (item_id) {
       case MENU_SAVE:
//         Toast.makeText(this, "您选中的是保存菜单", Toast.LENGTH_
SHORT).show();
           //在这里也可以跳转到其他的 Activity 中
           /*新建一个 Intent 对象*/
           Intent intent = new Intent();
           /*指定一个要启动的类*/
           intent.setClass(ActivityMenu.this, ActivityMain.class);
```

```
            /*启动一个新的 Activity*/
            startActivity(intent);
            /*关闭当前的 Activity*/
            ActivityMenu.this.finish();
            break;
        case MENU_DELETE:
            //在这里做相应的处理
            ActivityMenu.this.finish();
            break;
        }
        return true;
    }
```

4.5 列表（ListView）

4.5.1 列表（ListView）简介

前面介绍了界面布局、界面跳转，以及常用的组件，如 Button、TestView、EditText 等。本节将重点介绍 Android 的另一个非常重要的组件 ListView，也就是常说的列表。

图 4-13 所示就是一个简单的列表。4.5.2 节将讲解怎样创建一个 ListView。

图 4-13

4.5.2 简单 ListView 的创建方式

图 4-13 所示的 ListView 以列表的形式显示每条信息，并且每条信息都可以被选择，用

于执行相应的操作，创建 ListView 步骤如下所述。

第一步：创建一个新的名为：listViewDemo 的 Android 项目。

第二步：在 main.xml 文件中添加一个 Button，并在 ActivityMain 中获得此 Button。添加监听，然后单击该 Button，跳转到 ActivityList.class 中。

第三步：在 ActivityList.java 文件中编写代码如下所示：

```
package com.redarmy.list;

import android.app.Activity;
import android.os.Bundle;
import android.widget.ArrayAdapter;
import android.widget.ListView;
public class ActivityList01 extends Activity {
   public ListView listView;
   public String className[] = { "3G 数字内容学院 0216 班", "3G 数字内容
学院 0309 班",
          "3G 数字内容学院 0720 班", "3G 数字内容学院 0727 班", "3G 数字内容
学院 0105 班", "3G 数字内容学院 0302 班",
          "3G 数字内容学院 0503 班" };

   @Override
   protected void onCreate(Bundle savedInstanceState) {
      // TODO Auto-generated method stub
      super.onCreate(savedInstanceState);
      listView = new ListView(this);
//    listView.setAdapter(new ArrayAdapter<String>(this,
//          android.R.layout.simple_list_item_1, className));
//    setContentView(listView);

      listView.setAdapter(new          ArrayAdapter<String>(this,
android.R.layout.simple_list_item_single_choice,className));
      listView.setItemsCanFocus(true);
      listView.setChoiceMode(ListView.CHOICE_MODE_MULTIPLE);
      setContentView(listView);
   }

}
```

在上述代码中，ListView 没有任何内容，因此使用 ListView 的 setAdapter()方法来为其填充内容，代码如下：

```
listView.setAdapter(new ArrayAdapter<String>(this,
          android.R.layout.simple_list_item_1, className));
```

该语句让 listView 和 ArrayAdapter 绑定。ListView 是一个列表，用于显示数据。在本方法中，列表将 ArrayAdapter 的数据显示出来。ArrayAdapter 由三个参数构造。第一个参数是

Context，即上下文的引用；第二个参数是在 R 文件中定义的 Layout，只不过这里是系统的 R 文件；第三个参数是数组，其中每一项的类型没有限制。

该例需要一个 Adapter 类型参数。那么，什么是 Adapter 呢?为什么要使用 Adapter？4.5.3 节将详细介绍。

4.5.3 Adapter 接口

在 Android 平台上，不允许直接将字符串数组应用在界面组件中。界面菜单的项目都要由接口（Adapter）提供。为什么需要在界面与数据之间添加一个“接口”呢？答案是“为了保持程序的弹性”。因为除了 Android 提供的各种标准界面组件之外，用户还可以自定义各种界面组件。只要在界面组件与数据之间自定义一个接口，就可以在自定义界面组件中使用原本的数据源。

Android 平台默认提供的适配器类型有很多种。ArrayAdapter 的作用是读入程序中已声明的数组，并转换成界面组件能识别的组件。除此之外，界面组件可以识别的适配器还有 SimpleAdapter（从 XML 文字字符串文件读入数组）、CursorAdapter（从 ContentProvider 读入数组）等。

之所以要有多种适配器，是因为数据源的类型不同，但是界面组件接收数据的格式单一。就像生活中不同的电器有不同的供电规格，采用不同供电规格的电器不能使用相同的电源，需要用电源适配器，使市电与电器电源匹配。在 Android 平台上，Adapter 起到适配器的作用。不管原来的数据源（220V 市电）是什么，选择适当的适配器（电源适配器）来转换数据源，以便在规格单一的界面组件（市电插座）上使用。

所以，Adpater 接口的实现类起到桥梁的作用。图 4-14 所示是常用的实现 Adapter 接口的类，详述如下。

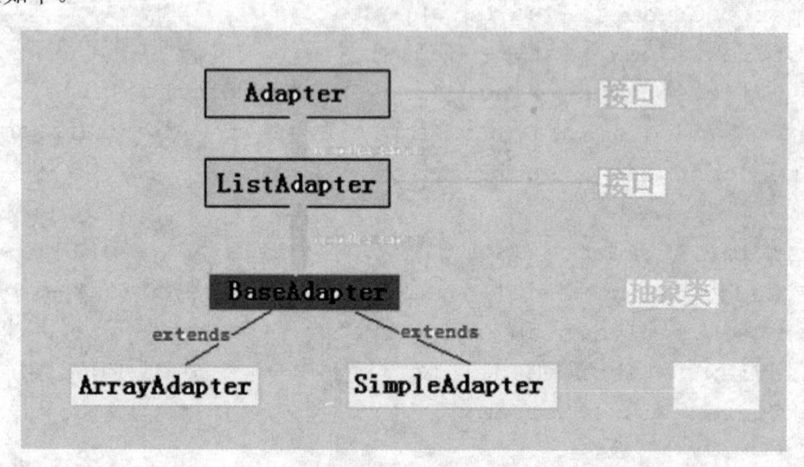

图 4-14

1. ListAdapter

ListAdapter 继承于 Adapter，是 ListView 及其数据的适配器。也就是说，要让 ListView 显示出来，需要以下三个条件：

（1）ListView：需要被显示的列表。

（2）Data：ListView 绑定的数据，一般是一个 Cursor 或者一个字符串数组。

（3）ListAdapter：是 data 和 listView 的桥梁，起到适配的作用。

2．ArrayAdapter

ArrayAdapter 是 ListAdapter 的直接子类，直译为数组适配器。显然，它是数组和 listView 之间的桥梁，将数组定义的数据一一对应地显示在 ListView 里。通常，ArrayAdapter 适配的 ListView 的每一项只有一个 TextView，其中显示的内容就是数组里对象调用 toString()方法生成的字符串。

通过以下方式，将 listView 和 ArrayAdapter 绑定：

```
listView.setAdapter(new ArrayAdapter<String>(this,
            android.R.layout.simple_list_item_1, className));
```

Android 系统默认的布局方式通过 Android.R.layout.XX 来定义，常见的有以下几种：

（1）android.R.Layout.simple_list_item_1，每项只有一个 TextView。

（2）android.R.Layout.simple_list_item_2，每项有两个 TextView。

（3）android.R.Layout.simple_list_item_single_choice，每项有一个 TextView，但是可选。

3．SimpleAdapter

SimpleAdapter 也是 ListAdatper 直接实现的一个类。通过 SimpleAdapter，可以让 ListView 中每一项的内容更加个性化。通常，ListView 中某项的布局信息写在 XML 的布局文件中，通过 R.layout.xx（xx 为文件名）获得；也可以使用系统自带的布局文件。

ArrayAdapter 是数组和 ListView 间的桥梁，SimpleAdapter 是 ArrayList 和 ListView 间的桥梁。需要注意的是，ArrayList 的每一项都是 Map<String,?>类型。ArrayList 的 Map 对象和 ListView 的每一项数据绑定，并且一一对应。

SimpleAdapter 的构造函数为：

```
public SimpleAdapter(Context context, List<? extends Map<String, ?>>
data, int resource, String[] from, int[] to)
```

其中：

（1）Context：很多类的构造器将此上下文应用传递进去。

（2）data：基于 Map 的 list。data 和 ListView 的每一项对应。data 项是 Map 类型，包含 ListView 每一行需要的数据。常用的用法是 data = new ArrayList<Map<String,Object>>();。

（3）resource：是一个 Layout，包含在 to 出现的 View 中。一般采用系统提供的资源，也可以自定义。

（4）from：是一个数组名称，ArrayList 的 item 索引 Map<String,Object>的 Object 时使用。

to 里是一个 TextView 数组，以 id 的形式表示，例如 android.R.id.text1，text1 在 layout 中可以索引到。需要注意的是，这些 View 必须是 TextView。

很多时候不仅要求显示列表，还需要和列表交互，例如，单击列表中的某一项，触发相应的动作。列表的交互是经常的，有时候也是必须的。list 的事件处理是把 OnItemClickListener 注册到 listView 中，如下所示：

```
listView.setOnItemClickListener(new OnItemClickListener() {
   @Override
   public void onItemClick(AdapterView<?> arg0, View arg1, int arg2,
         long arg3) {
      setTitle(arg0.getItemAtPosition(arg2).toString());
   }
});
```

或者通过 OnItemClickListener 接口的 onItemClick（AdapterView<?> parent, View view, int position, long id）方法，完成单击 listView 每一项的事件处理。

另外，选择事件的接口 AdapterView.OnItemSelectedListener 有两个方法，如下所示：

（1）abstract void onItemSelected(AdapterView<?> parent, View view, int position, long id)

（2）abstract void onNothingSelected(AdapterView<?> parent)

其实现方式与 OnItemClickListener 相同。

注意：很多情况下，每一个 Activity 只有一个 ListView，而且会占满一屏。所以，Android 提供了一个类，叫做 ListActivity。它既是 Activity，又有 ListView 的特征。

每个 ListActivity 都和一个布局文件相关联，而且布局文件按照下列模式固定的：

```
<?xml version="1.0" encoding="utf-8"?>
 <LinearLayout
xmlns:android="http://schemas.android.com/apk/res/android"
       android:orientation="vertical"
       android:layout_width="fill_parent"
       android:layout_height="fill_parent"
       android:paddingLeft="8dp"
       android:paddingRight="8dp">

    <ListView android:id="@id/android:list"
            android:layout_width="fill_parent"
            android:layout_height="fill_parent"
            android:background="#00FF00"
            android:layout_weight="1"
            android:drawSelectorOnTop="false"/>

    <TextView id="@id/android:empty"
            android:layout_width="fill_parent"
            android:layout_height="fill_parent"
            android:background="#FF0000"
            android:text="No data"/>
 </LinearLayout>
```

Row Layout 的布局文件如下所示：

```
<?xml version="1.0" encoding="utf-8"?>
 <LinearLayout
xmlns:android="http://schemas.android.com/apk/res/android"
    android:layout_width="fill_parent"
    android:layout_height="wrap_content"
    android:orientation="vertical">

    <TextView android:id="@+id/text1"
       android:textSize="16sp"
       android:textStyle="bold"
```

```
    android:layout_width="fill_parent"
    android:layout_height="wrap_content"/>

  <TextView android:id="@+id/text2"
    android:textSize="16sp"
    android:layout_width="fill_parent"
    android:layout_height="wrap_content"/>
</LinearLayout>
```

或者采用系统提供的布局文件。

有关 ListView 使用的详细信息，请查阅 Android 提供的 API 或者相关文档。

在学习过程中，还可以参考 listDemo 案例。

4.6　对话框（Dialog）

4.6.1　对话框（Dialog）简介

对话框是一种显示在 Activity 主界面上的用户界面元素。虽然对话框在程序中不是必需的，但是当 Activity 需要临时显示一些信息或提供一些功能，而新建 Activity 的开销太大时，采用对话框不仅可以节省开销，也可以增强应用的友好性。如今崇尚用户体验至上，好的界面，意味着成功了一半，所以要学好对话框的应用，并在适当的地方灵活应用。常用的场景，如用户登录、网络下载、下载成功或失败、来短信、电量不足等，都需要对话框提示。

Android 系统提供的常用 Dialog 类及其继承关系如图 4-15 所示。

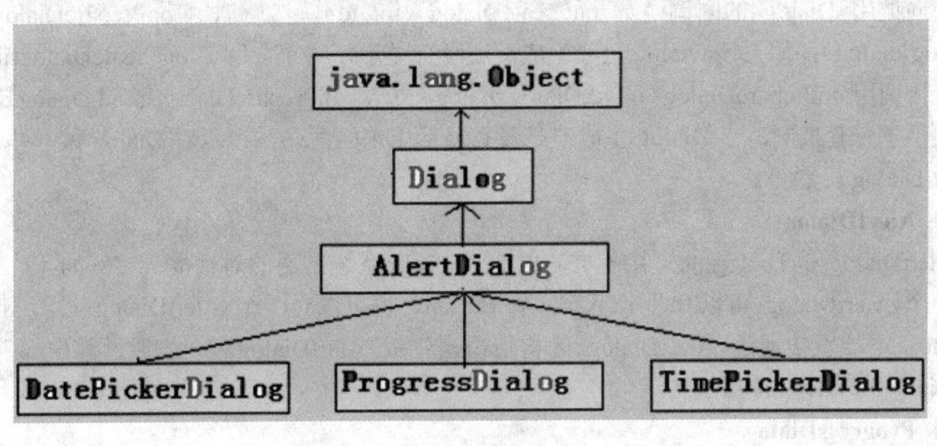

图　4-15

图中所示的 Dialog 类实现了一个非常重要的接口：DialogInterface，其中封装了一些很重要的监听，请参考 Android 的帮助文档 API。下面简单介绍几个常用的类。

1.　Dialog

Dialog 类是一切对话框的基类。需要注意的是，Dialog 类虽然可以在界面上显示，但是并非继承于常用的 View 类，而是直接从 java.lang.Object 开始构造。类似于 Activity，Dialog 也是有生命周期的，由 Activity 维护。Activity 负责生成、保存、恢复它。在生命周期的每个阶段都有一些回调函数供系统反向调用。Dialog 的生命周期如图 4-16 所示，简述如下。

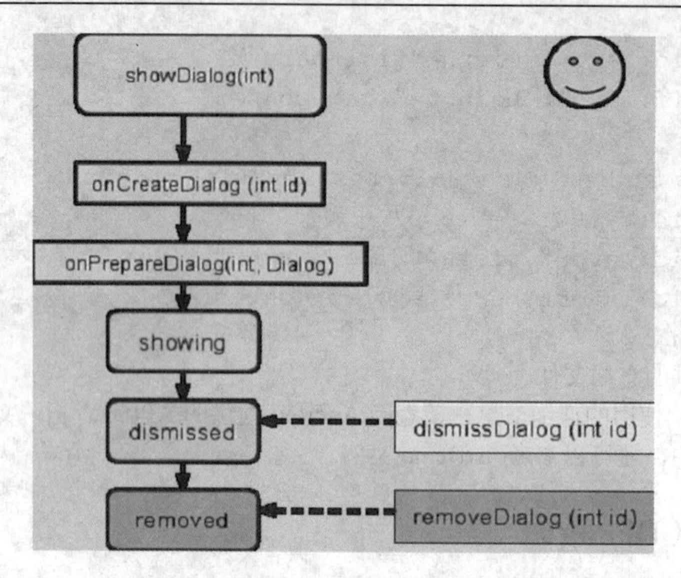

图 4-16

在 Activity 中，用户可以主动调用的函数是：

（1）showDialog(int id)，负责显示标识为 id 的 Dialog。函数调用后，系统将反向调用 Dialog 的回调函数 onCreateDialog(int id)。

（2）dismissDialog，使标识为 id 的 Dialog 在界面中消失。

Dialog 有两个比较常见的回调函数：oncreateDialog（int id）和 onPrepareDialog（int id,Dialog dialog）。在 Activity 中调用 showDialog（int id），如果该 Dialog 是第一次生成，系统将反向调用 Dialog 的回调函数 onCreateDialog（int id），然后调用 onPrepareDialog（int id,Dialog dialog）；如果该 Dialog 已经生成，还没有显示，将不会回调 onCreateDialog(int id), 而是直接调用 onPrepareDialog（int id,Dialog dialog）方法。onPrepareDialog(int id,Dialog dialog) 方法提供了一套机制，当 Dialog 生成但是没有显示出来的时候，可以对 Dialog 做一些修改，如修改 Dialog 标题等。

2．AlertDialog

AlertDialog 是 Dialog 的直接子类，也是 Android 系统常用的对话框，实例如 4.6.2 节所述。一个 AlertDialog 可以有两个或者三个 Button，也可以对一个 AlertDialog 设置 title 和 message。不能直接通过 AlertDialog 的构造器来生成 AlertDialog，一般通过其内部静态类 AlertDialog.Builder 来构造。

3．ProgressDialog

ProgressDialog 负责显示进度的相关情况，它是 AlertDialog 的一个子类。在 4.6.2 节所述实例中，实现默认的进度条。也可以配置自己的进度条。与 AlertDialog 不同，ProgressDialog 不需要通过 Builder 这个内部静态类，而是直接使用构造函数来构造。

4．DatePickerDialog 和 TimePickerDialog

DatePickerDialog 和 TimePickerDialog 是日期选择对话框和时间选择对话框，分别用于设置日期和时间。

Android 中有许多 Dialog，最常用的是 AlertDialog 和 ProgressDialog。对于其他 Dialog，请参考 Android 的 API 文档和实例。还可以创建自定义对话框。

4.6.2　创建 AlertDialog 解析常用的对话框方法

AlertDialog 在对话框中经常使用，本节通过一个实例详细介绍其用法。

第一步：创建一个名为 AlertDialogDemo 的 Android 项目。

第二步：在 main.xml 文件中添加一个 Button 按钮，单击后显示 AlertDialog，代码如下所示：

```
<?xml version="1.0" encoding="utf-8"?>
<LinearLayout
xmlns:android="http://schemas.android.com/apk/res/android"
    android:orientation="vertical"
    android:layout_width="fill_parent"
    android:layout_height="fill_parent"
    >
<Button android:id="@+id/Button01" android:layout_height="wrap_content"
android:text="@string/AlertDialog"
android:layout_width="fill_parent"></Button>
</LinearLayout>
```

第三步：在 strings.xml 文件中添加要用到的字符串，在 drawable 中添加要用到的图片。

第四步：在类中获得 Button 实例并添加监听，并重写 onCreateDialog（int id）的方法。

```
public class AlertDialogSamples extends Activity implements
OnClickListener {
    private static final int DIALOG_YES_NO_MESSAGE = 1;
    /** Called when the activity is first created. */
    @Override
    public void onCreate(Bundle savedInstanceState) {
        super.onCreate(savedInstanceState);
        setContentView(R.layout.main);
        // 得到组件
        Button btn = (Button) this.findViewById(R.id.Button01);
        // 添加监听
        btn.setOnClickListener(this);
    }
    @Override
    public void onClick(View v) {
        int temp = v.getId();
        switch (temp) {
        case R.id.Button01:
            showDialog(DIALOG_YES_NO_MESSAGE);
            break;
        }
    }
    @Override
    protected Dialog onCreateDialog(int id) {
```

```
        switch (id) {
        case DIALOG_YES_NO_MESSAGE:
            return new AlertDialog.Builder(AlertDialogSamples.this)
            .setIcon(R.drawable.alert_dialog_icon)
            .setTitle(R.string.alert_dialog_two_buttons_title)
            .setPositiveButton(R.string.alert_dialog_ok, new
DialogInterface.OnClickListener() {
                @Override
                public void onClick(DialogInterface dialog, int which) {

                }
            })
            .setNegativeButton(R.string.alert_dialog_cancel, new
DialogInterface.OnClickListener() {
                @Override
                public void onClick(DialogInterface dialog, int which) {
                    // TODO Auto-generated method stub

                }
            }).create();
        }
        return null;
    }
}
```

这里用到了 AlertDialog。AlertDialog 是最基本的对话框,它支持在对话框底部显示最多三个按钮，并根据单击的按钮执行不同的操作。AlertDialog.Builder 提供了更方便的初始化对话框的方法，如表 4-2 所示。

表 4-2

方法名	描述
setIcon	设置标题图标
setTitle	设置标题文本
setCustomTitle	传入一个 View 对象作为自定义标题，并使 setIcon 和 setTitle 设置的标题失效
setMessage	设置对话框内容为传入的字符串
setItems setSingleChoiceItems setMultiChoiceItems	设置对话框的内容为一个列表。需要输入一个字符串数组的资源 id。字符串数组资源定义在 res/values/array.xml 中。定义字符串数组的示例如下： <resources> <string-array name="*myArray*"> <item>3G 数字内容学院</item> <item>东方尚智教育科技有限公司</item> </string-array> </resources> 另外，需要接入一个监听器，处理列表元素被单击的事件。setSingleChoiceItems 设置的内容为单选列表,setMultiChoiceItems 设置的内容为复选列表

续表

方　法　名	描　　述
setView	设置对话框的内容为自定义 View
setPositiveButton setNeutralButton setNegativeButton	设置对话框底部的按钮，然后显示。输入一个字符串，并设置显示的文本，接入一个监听器，处理按钮的单击事件。三个按钮的位置从左到右分别为确定（Positive）、中立（Neutral）和否定（Negative）

4.6.3　创建对话框（Dialog）

创建 Dialog 可采用以下方式：

（1）继承 Dialog 的类。

（2）重写 onCreateDialog(int id)的方法：

① setTitle()：设置标题；

② setContentView()：设置显示内容。

具体操作与在 Java 中创建 GUI 窗口类似。

4.6.4　对话框（Dialog）应用实例

本节详细分析一个登录的实例，进一步探讨 Dialog。

第一步：创建 Android 项目，名为：LoginDialogDemo。

第二步：在 main.xml 文件添加 Button（按钮），代码如下所示：

```
<?xml version="1.0" encoding="utf-8"?>
<LinearLayout
xmlns:android="http://schemas.android.com/apk/res/android"
   android:orientation="vertical" android:layout_width="fill_parent"
   android:layout_height="fill_parent">

   <Button android:id="@+id/Button01"android:layout_width="wrap_content"
      android:layout_height="wrap_content"
android:text="@string/StartDialog"></Button>
</LinearLayout>
```

第三步：增加登录的 login_dialog.xml 文件，代码如下所示：

```
<LinearLayout android:id="@+id/LinearLayout01"
   android:layout_width="fill_parent"
android:layout_height="fill_parent"
   xmlns:android="http://schemas.android.com/apk/res/android"
   android:orientation="vertical">
   <TextView android:layout_width="wrap_content"
      android:layout_height="wrap_content"
android:id="@+id/username"
      android:text="@string/username"
android:layout_marginLeft="20dip"
      android:layout_marginRight="20dip"
android:layout_gravity="left"></TextView>
```

```
    <EditText android:layout_height="wrap_content"
 android:id="@+id/username"
        android:layout_width="fill_parent"
android:layout_marginLeft="20dip"
        android:layout_marginRight="20dip"
android:layout_gravity="fill_horizontal"
        android:capitalize="none" android:autoText="false"></EditText>
    <TextView android:layout_width="wrap_content"
        android:layout_height="wrap_content"
android:id="@+id/password"
        android:text="@string/userpass"
android:layout_marginLeft="20dip"
        android:layout_marginRight="20dip"
android:gravity="left"></TextView>
    <EditText android:text="EditText01"
android:layout_height="wrap_content"
        android:layout_marginLeft="20dip"
android:layout_marginRight="20dip"
        android:layout_width="fill_parent"
android:layout_gravity="fill_horizontal"
        android:id="@+id/password" android:capitalize="none"
android:autoText="false"
        android:password="true"></EditText>
</LinearLayout>
```

第四步：在 res/drawable 目录下增加 alert_dialog_icon.png，代码如下所示：

```
<?xml version="1.0" encoding="utf-8"?>
<resources>
    <string name="userpass">密码</string>
    <string name="username">账号</string>
    <string name="StartDialog">Start 开始 Dialog 的演示</string>
    <string name="hello">Hello World, ActivityMain!</string>
    <string name="app_name">LoginDialog</string>
</resources>
```

第五步：在 ActivityMain 中实现，代码如下所示：

```
package com.redarmy.login;

import android.app.Activity;
import android.app.AlertDialog;
import android.app.Dialog;
import android.app.ProgressDialog;
import android.content.DialogInterface;
```

```
import android.os.Bundle;
import android.view.LayoutInflater;
import android.view.View;
import android.view.View.OnClickListener;
import android.widget.Button;

public class ActivityMain extends Activity {
    public static final int DIALOG_SHOW = 1;
    public ProgressDialog m_Dialog;
    @Override
    public void onCreate(Bundle savedInstanceState) {
        super.onCreate(savedInstanceState);
        setContentView(R.layout.main);
        Button btn = (Button) findViewById(R.id.Button01);
        btn.setOnClickListener(new OnClickListener() {
            @Override
            public void onClick(View v) {
                showDialog(DIALOG_SHOW);
            }
        });
    }
    @Override
    protected Dialog onCreateDialog(int id) {
        switch (id) {
        case DIALOG_SHOW:
            return new AlertDialog.Builder(ActivityMain.this)
            .setIcon(R.drawable.alert_dialog_icon)    //设置图片
                   .setTitle("登录提示")               //设置标题
                   .setMessage("这里需要登录")          //设置显示信息
                   .setPositiveButton("确定")          //设置按钮
                  new DialogInterface.OnClickListener() {
                      @Override
                      public void onClick(DialogInterface dialog, int
which) {
                          //单击“确定”转向登录框
                          //第一步：得到一个 LayoutInflater 对象，对 xml 布
局进行解析，并生成一个 View
                         LayoutInflater inflater = LayoutInflater.from
(ActivityMain.this);
                         //得到自定义对话框
                         View dialogView = inflater.inflate(R.layout.
dialog_show, null);
```

```
                    Dialog dlg = new
AlertDialog.Builder(ActivityMain.this)
                    .setTitle("登录框")
                    .setView(dialogView)//设置自定义View
                    .setPositiveButton("确定"), new
DialogInterface.OnClickListener() {
                     @Override
                     public void onClick(DialogInterface dialog,
int which) {
                        m_Dialog =ProgressDialog.show(ActivityMain.
this, "请等待...", "正在为登录",true);

                        //创建线程的匿名类,3 秒钟,登录不成功
                        new Thread(){
                           public void run(){
                              try{
                              Thread.sleep(1000*3);
                              }catch(Exception ex){
                              }finally{
                                 //登录结束,取消登录对话框
                                 m_Dialog.dismiss();
                              }
                           }
                        }.start();
                     }
                  }).setNegativeButton("取消", new
DialogInterface.OnClickListener() {
                     @Override
                     public void onClick(DialogInterface dialog,
int which) {
                        ActivityMain.this.finish();
                     }
                  }).create();
                 dlg.show();//显示
                  }
               }).setNeutralButton("退出",
               new DialogInterface.OnClickListener() {
                  @Override
                  public void onClick(DialogInterface dialog, int
which) {
                     ActivityMain.this.finish();
                  }
```

```
                    }).create();
        }
        return null;
    }
}
```

4.7　Toast 和 Notification 的应用

对话框（Dialog）其实起到了提醒的作用，但是 Android 系统提供了一套更加友好的、轻量级的机制来提醒用户。这种系统不会打断用户的当前操作,设计得非常巧妙。

4.7.1　Toast

Toast 是 Android 提供的快显讯息类，是轻量级的提醒机制。Toast 类的使用非常简单，而且用途广泛。Toast 永远不获得聚焦，不会打断用户当前的操作，其信息在 floating view 呈现，然后自动消失。

Toast 能够更加友好地提示用户。可采用以下两种方法创建 Toast。

1．简单文字信息

（1）通过 make()方法创建 Toast 信息。

（2）调用 show()方法显示 Toast 提示信息。

2．复杂 Toast 信息

Toast 支持通过 setView(view)添加 view 组件。

4.7.2　Notification

当有未接电话或短信时，在 Android 手机的顶部状态栏将出现一个小图标，提示用户有未处理的快讯。用触笔按住状态栏往下拖动，就可以展开并查看这些快讯。Android 平台提供了 NotificationManager 来管理状态栏信息，用 Notification 来处理信息。Notification 是 Android 提供的在状态栏的提醒机制，它同样不会打断用户当前的操作，并且支持更复杂的单击事件响应。

创建 Notification 的四个步骤如下所述。

（1）得到一个 NotificationManager 的引用。

```
String ns = Context.NOTIFICATION_SERVICE;
NotificationManager nManager=(NotificationManager)
getSystemService(ns);
```

（2）初始化一个 Notification。

```
int icon = R.drawable.notification_icon;
CharSequence tickerText="Hello";
long when = System.currentTimeMillis();
Notification notification = new Notification(icon,tickerText,when);
```

（3）设置 Notification 参数。

```
Context context = getApplicationContext();
CharSequence contentTitle="My notification";
CharSequence contentText="Hello World";
Intent notificationIntent = newt Intent(this,MyClass.class);
```

```
PendingIntent    pIntent    =    PendingIntent.getActivity(this,0,
notificationintent,0);
notification.setLatestEventInfo(context,contentTitle,contentText,p
Intent);
```

（4）显示一个 Notification。

```
private static final int HELLO_ID=1;
nManager.notify(HELLO_ID,notification);
```

4.7.3 Toast 与 Notification 应用实例

Toast 与 Notification 的实例详述如下。

第一步：创建 Android 项目，名为 ToastNotDemo。

第二步：在 res/drawable 下添加图片，并修改 main.xml 配置文件及 res/value/strings 文件 main.xml，代码如下所示：

```
<?xml version="1.0" encoding="utf-8"?>
<LinearLayout
xmlns:android="http://schemas.android.com/apk/res/android"
   android:orientation="vertical" android:layout_width="fill_parent"
   android:layout_height="fill_parent">

   <Button android:layout_height="wrap_content"
 android:id="@+id/ToastBtn"
      android:layout_width="fill_parent"
android:layout_marginLeft="20dip"
      android:layout_marginRight="20dip"
android:text="@string/ToastBtn"></Button>
   <Button android:layout_height="wrap_content"
 android:id="@+id/NotBtn"
      android:text="@string/NotBtn"
android:layout_width="fill_parent"
      android:layout_marginLeft="20dip"
android:layout_marginRight="20dip"></Button>
</LinearLayout>
```

res/values/strings 的代码如下所示：

```
<?xml version="1.0" encoding="utf-8"?>
<resources>
   <string name="NotTextView">高级的 Notification</string>
   <string name="clearBtn">清除 Notification</string>
   <string name="soundandshock">带有声音和振动混合的 Notification
</string>
   <string name="shockBtn">带有振动的 Notification</string>
   <string name="soundBtn">带有声音的 Notification</string>
```

```
    <string name="rainBtn">大雨连绵</string>
    <string name="cloudyBtn">阴云密布</string>
    <string name="sunBtn">晴空万里</string>
    <string name="NotBtn">Notification 演示案例</string>
    <string name="ToastBtn">Toast 演示案例</string>
    <string name="hello">Hello World, ActivityMain!</string>
    <string name="app_name">ToastAndNot</string>

     <string name="LongToastBtn">长时间 Toast 演示案例</string>
    <string name="ShortToastBtn">短时间 Toast 演示案例</string>
</resources>
```

第三步：新建 ActivityToast 类、ActivityNot 类及其配置文件。

ActivityToast 代码如下所示：

```
package com.redarmy.ToastAndNot;

import android.app.Activity;
import android.app.Notification;
import android.app.NotificationManager;
import android.app.PendingIntent;
import android.content.Context;
import android.content.Intent;
import android.os.Bundle;
import android.view.LayoutInflater;
import android.view.View;
import android.view.View.OnClickListener;
import android.widget.Button;
import android.widget.TextView;
import android.widget.Toast;

public class ActivityToast extends Activity implements OnClickListener
{
    public static final int NOTIFICATIONS_ID=R.layout.toast;

    public void onCreate(Bundle savedInstanceState) {
        super.onCreate(savedInstanceState);
        setContentView(R.layout.toast);
        // 获取组件对象
        Button ShortToastBtn = (Button) findViewById(R.id.ShortToastBtn);
        Button LongToastBtn = (Button) findViewById(R.id.LongToastBtn);
        // 添加监听
        ShortToastBtn.setOnClickListener(this);
        LongToastBtn.setOnClickListener(this);
```

```
    }
    @Override
    public void onClick(View v) {
        int temp = v.getId();
        switch (temp) {
        case R.id.ShortToastBtn:
            setTitle("短时间演示案例");
            showToast(Toast.LENGTH_SHORT);
            break;
        case R.id.LongToastBtn:
            setTitle("长时间演示案例");
            showToast(Toast.LENGTH_LONG);
            showNotification();
            break;
        }
    }

    public View inflateView(int resource) {
        LayoutInflater lvi = (LayoutInflater)
 getSystemService(Context.LAYOUT_INFLATER_SERVICE);
        return lvi.inflate(resource, null);
    }

    public void showToast(int type) {
        View view = inflateView(R.layout.toast_show);
        TextView tv = (TextView) view.findViewById(R.id.content);
        tv.setText("欢迎您来到东方尚智 3G 数字内容学院,这是位于北京的 3G 培训学
院.");
        Toast toast = new Toast(this);
        toast.setView(view);
        toast.setDuration(type);
        toast.show();
    }

    public void showNotification() {

      NotificationManager notificationManager = (NotificationManager)
getSystemService(NOTIFICATION_SERVICE);

        CharSequence title = "最专业的 Android 培训学院";
        CharSequence contents = "3gdci.com";
```

```
        PendingIntent contentIntent = PendingIntent.getActivity(this, 0,
                new Intent(this, ActivityMain.class), 0);

        Notification notification = new
Notification(R.drawable.default_icon,
                title, System.currentTimeMillis());
        notification.setLatestEventInfo(this,    title,    contents,
contentIntent);

        // 100ms 延迟后，振动 250ms；停止 100ms 后，振动 500ms
        notification.vibrate = new long[] { 100, 250, 100, 500 };
        notificationManager.notify(NOTIFICATIONS_ID, notification);
    }

}
```

配置文件 toast.xml 的代码如下所示：

```
<?xml version="1.0" encoding="utf-8"?>
<LinearLayout
xmlns:android="http://schemas.android.com/apk/res/android"
    android:orientation="vertical" android:layout_width="fill_parent"
    android:layout_height="fill_parent">
    <Button android:layout_height="wrap_content"
 android:id="@+id/ShortToastBtn"
        android:layout_width="fill_parent"
android:layout_marginLeft="20dip"
        android:layout_marginRight="20dip"
android:text="@string/ShortToastBtn"></Button>
    <Button android:layout_height="wrap_content"
 android:id="@+id/LongToastBtn"
        android:text="@string/LongToastBtn"
android:layout_width="fill_parent"
        android:layout_marginLeft="20dip"
android:layout_marginRight="20dip"></Button>
</LinearLayout>
```

配置文件 toast_show.xml 的代码如下所示：

```
<FrameLayout android:id="@+id/FrameLayout01"
    android:layout_width="fill_parent"
    xmlns:android="http://schemas.android.com/apk/res/android"
    android:layout_height="wrap_content"
    android:background="@android:drawable/toast_frame">

    <LinearLayout android:id="@+id/LinearLayout01"
```

```
        android:layout_height="wrap_content"
        android:layout_width="fill_parent"
        android:orientation="horizontal">

        <ImageView android:id="@+id/ImageView01"
            android:layout_width="wrap_content"
            android:layout_height="wrap_content"
            android:src="@drawable/default_icon">
            </ImageView>
        <TextView android:id="@+id/content"
                android:layout_width="wrap_content"
                android:layout_height="wrap_content"
                android:paddingLeft="6dip"
                android:gravity="center_vertical">
            </TextView>
    </LinearLayout>
</FrameLayout>
```

ActivtiyNot 类的代码如下所示：

```
package com.redarmy.ToastAndNot;

import android.app.Activity;
import android.app.Notification;
import android.app.NotificationManager;
import android.app.PendingIntent;
import android.content.Intent;
import android.os.Bundle;
import android.view.View;
import android.view.View.OnClickListener;
import android.widget.Button;

public class ActivityNot extends Activity implements OnClickListener
{

    public NotificationManager mNotM;
    public static final int NOTID=R.layout.notification_show;

    public void onCreate(Bundle savedInstanceState) {
        super.onCreate(savedInstanceState);
        setContentView(R.layout.notification_show);
        // 获取第一个 LinearLayout 组件
        Button SunBtn = (Button) findViewById(R.id.SunBtn);// 晴天
        Button cloudyBtn = (Button) findViewById(R.id.cloudyBtn);//阴云
```

```
        Button rainBtn = (Button) findViewById(R.id.rainBtn);//大雨

        // 获取第二个 LinearLayout 组件
        Button soundBtn = (Button) findViewById(R.id.soundBtn);//声音
        Button shockBtn = (Button) findViewById(R.id.shockBtn);//振动
        Button soundandshockBtn = (Button)
findViewById(R.id.soundandshock);// 既有声音，又有振动

        // 绑定所有组件
        SunBtn.setOnClickListener(this);
        cloudyBtn.setOnClickListener(this);
        rainBtn.setOnClickListener(this);
        // 第二部分
        soundBtn.setOnClickListener(this);
        shockBtn.setOnClickListener(this);
        soundandshockBtn.setOnClickListener(this);
        //第一步：得到 NotificationManager 的一个引用
        mNotM =(NotificationManager)getSystemService (NOTIFICATION_
SERVICE);
    }

    @Override
    public void onClick(View v) {
        int temp = v.getId();
        switch (temp) {
        case R.id.SunBtn:
            setWeather("晴空万里", "天气预报", "晴空万里", R.drawable.sun);
            break;
        case R.id.cloudyBtn:
            setWeather("阴云密布","天气预报","阴云密布",R.drawable.cloudy);
            break;
        case R.id.rainBtn:
            setWeather("大雨连绵", "天气预报", "大雨连绵", R.drawable.rain);
            break;
        case R.id.soundBtn:
            setDefault(Notification.DEFAULT_SOUND);       //声音
            break;
        case R.id.shockBtn:
            setDefault(Notification.DEFAULT_VIBRATE);    //振动
            break;
        case R.id.soundandshock:
            setDefault(Notification.DEFAULT_ALL);         //振动与声音
```

```
            break;
        }

    }

    private void setDefault(int defaults) {
        //第二步：初始化一个Notification
        int icon = R.drawable.sun;
        CharSequence tickerText="晴空万里";
        long when = System.currentTimeMillis();//得到系统当前时间
        //构造器参数的含义：1.显示的图片id;2.显示的文本;3.时间
        Notification notification = new
 Notification(icon,tickerText,when);
        //第三步:设置Notification的参数
        Intent notIntent = new Intent(this,ActivityMain.class);
        PendingIntent pIntent = PendingIntent.getActivity(this, 0,
notIntent, 0);

        String title="晴空万里";
        notification.setLatestEventInfo(this, title, tickerText,
pIntent);
        //第四步:显示
        notification.defaults=defaults;

        mNotM.notify(NOTID, notification);

    }

    private void setWeather(String tickerText, String title, String
content,
            int drawable) {
        //第二步：初始化一个Notification
        long when = System.currentTimeMillis();//得到系统当前时间
        //构造器参数的含义：1.显示的图片id;2.显示的文本;3.时间
        Notification notification = new
 Notification(drawable,tickerText,when);
        //第三步:设置Notification的参数
        Intent notIntent = new Intent(this,ActivityMain.class);
        PendingIntent pIntent = PendingIntent.getActivity(this, 0,
notIntent, 0);
```

```
        notification.setLatestEventInfo(this, title, content,
pIntent);
        //第四步：显示
        mNotM.notify(NOTID, notification);
    }
}
```

配置文件 notification_show.xml 的代码如下所示：

```
<ScrollView android:id="@+id/ScrollView01"
    android:layout_width="fill_parent"
android:layout_height="fill_parent"
    xmlns:android="http://schemas.android.com/apk/res/android">
    <LinearLayout android:id="@+id/LinearLayout01"
        android:layout_height="wrap_content"
android:orientation="vertical"
        android:layout_width="fill_parent">
        <LinearLayout android:id="@+id/LinearLayout02"
            android:layout_height="wrap_content"
android:layout_width="fill_parent"
            android:orientation="vertical">
            <Button android:layout_height="wrap_content"
 android:id="@+id/SunBtn"
                android:text="@string/sunBtn"
android:layout_width="fill_parent"
                android:layout_marginLeft="20dip"
android:layout_marginRight="20dip"
                android:gravity="center"></Button>
            <Button android:layout_height="wrap_content"
 android:id="@+id/cloudyBtn"
                android:layout_width="fill_parent"
android:layout_marginLeft="20dip"
                android:layout_marginRight="20dip"
android:text="@string/cloudyBtn"></Button>
            <Button android:layout_height="wrap_content"
 android:id="@+id/rainBtn"
                android:text="@string/rainBtn"
android:layout_width="fill_parent"
                android:layout_marginLeft="20dip"
android:layout_marginRight="20dip"></Button>
        </LinearLayout>
        <TextView android:id="@+id/TextView01"
 android:layout_width="wrap_content"
```

```
            android:layout_height="wrap_content"
android:text="@string/NotTextView"
            android:layout_marginTop="20dip"></TextView>
        <LinearLayout android:id="@+id/LinearLayout03"
            android:layout_height="wrap_content"
android:layout_width="fill_parent"
            android:orientation="vertical">
            <Button android:layout_height="wrap_content"
 android:id="@+id/soundBtn"
                android:text="@string/soundBtn"
android:layout_width="fill_parent"
                android:layout_marginRight="20dip"
android:layout_marginLeft="20dip"></Button>
            <Button android:layout_height="wrap_content"
 android:id="@+id/shockBtn"
                android:text="@string/shockBtn"
android:layout_width="fill_parent"
                android:layout_marginLeft="20dip"
android:layout_marginRight="20dip"></Button>
            <Button android:layout_height="wrap_content"
 android:id="@+id/soundandshock"
                android:text="@string/soundandshock"
android:layout_width="fill_parent"
                android:layout_marginRight="20dip"
android:layout_marginLeft="20dip"></Button>
        </LinearLayout>

        <Button android:layout_height="wrap_content"
 android:id="@+id/clearBtn"
            android:text="@string/clearBtn"
android:layout_width="fill_parent"
            android:layout_marginRight="20dip"
android:layout_marginLeft="20dip"
            android:layout_marginTop="20dip"></Button>
    </LinearLayout>
</ScrollView>
```

第 5 章　Intent 和 Broadcast 应用

5.1　Intent 简介

Intent 可以说是 Android 的灵魂，程序跳转和传递数据基本上依靠 Intent。本章将详细讲解 Intent。在涉及异步操作或者需要做通知的时候，Android 提供了广播机制，本章将深入讲解。

5.1.1　Intent 基础

Intent 在 Android 应用中相当重要，对应用编程很有帮助。在 Android 的官方 API 中，对 Intent 是这样定义的：“An intent is an abstract description of an operation to be performed. It can be used with startActivity to launch an Activity, broadcastIntent to send it to any interested BroadcastReceiver components, and startService(Intent) or bindService(Intent, ServiceConnection, int) to communicate with a background Service.”也就是说，Intent 是一次对将要执行的操作的抽象描述，它有三种形式。Android 的三个基本组件 Activity、Service 和 BroadcastReceiver 都是通过 Intent 机制激活的，而不同类型的组件有传递 Intent 的不同方式。

（1）要激活一个新的 Activity，或者让现有的 Activity 执行新的操作，应调用 Context.startActivity()或 Activity.startActivityForResult()方法。这两个方法需要传入的 Intent 参数也称为 Activity Action Intent（活动行为意图）。根据 Intent 对象对目标 Activity 的描述，启动匹配的 Activity 或传递信息。

（2）要启动一个新的服务，或者向一个已有的服务传递新的指令，调用 Context.start Service()或 Context.bindService()方法，将此方法的上下文对象与 Service 绑定。

（3）通过 Context.sendBroadcast()、Context.sendOrderBroadcast()和 Context.sendStick Broadcast()方法发送 BroadcastIntent，所有已注册的拥有与之相匹配的 IntentFilter 的 BroadcastReceiver 被激活。这种机制广泛运用于设备或系统状态变化的通知。一个常见的例子是当 Android 的电池电量过低时，系统发送 Action 为 BATTERY_LOW 的广播，任何与该 Action 匹配的 IntentFilter 注册的 BroadcaseReceiver 都会运行自定义的处理代码，比如关闭设备的 Wi-Fi 和 GPS 以减少电池消耗。

Intent 一旦发出，Android 都会准确找到相匹配的一个或多个 Activity、Service 或 BroadcastReceiver 作为响应。所以，不同类型的 Intent 消息不会出现重叠：BrodcastIntent 消息只发送给 BroadcastReceiver，绝不可能发送给 Activity 或 Service。由 startActivity()传递的消息只可能发送给 Activity，由 startService()传递的 Intent 只可能发送给 Service。

广播（Broadcast）和服务（Service）的部分将在后面详细介绍。本节介绍通过 startActivity 方法启动新的 Activity。

5.1.2　用 Intent 启动新的 Activity

Intent 的用途是连接应用的各个 Activity。如果把 Activity 比作积木，Intent 就好像是胶水，把积木粘起来，构成建筑（如房子）。在程序中，如果要启动一个 Activity，通常调用 startActivity()方法，并把 Intent 作为参数传递下去，代码为：startActivity(myIntent)。

这个 MyIntent 或者指定了一个 Activity，或者包含了选定 Acitivity 的信息，具体启动哪个 Activity，由系统决定。Android 系统负责选择一个最满足匹配条件的 Activity。

1．启动特定的 Activity

启动特定 Activity 的代码如下所示：

```
Intent intent = new Intent(ActivityMain.this, OtherActivity.class);//
第一个参数是当前的 Context，第二个参数是跳转到新的 Activity
startActivity(intent);
```

执行 startActivity()方法后，新的 Activity 被创建(这里指 OtherActivity)，并且移到整个 Activity 的堆栈顶部。

2．启动未指明的 Activity

在程序当中，很多时候并不关心启动哪个 Activity，而是将想启动哪个 Activity 的描述信息置于 Intent 当中，然后由系统寻找匹配的 Activity 来启动。比如第三方 Activity，它只需要描述自己在什么情况下被执行，如果启动 Activity 的描述信息正好匹配第三方 Activity 的描述信息，那么第三方 Activity 被启动。例如：

```
Intent intent = new Intent(Intent.ACTION_DIAL,Uri.parse("tel:115-
1345"));
startActivity(intent);
```

上述代码没有指定 Activity，只是把启动的 Activity 的描述信息放在 Intent 中。执行 startActivity()方法后，Android 系统将寻找合适的 Activity，根据描述信息找到处理电话的 Activity，然后启动、执行。

3．处理 Activity 的返回值

在前面介绍用 startActivity()方法启动 Activity 时，没有传递数据，若需要传递数据，Android 系统提供了相应的处理机制。

startActivityForResult()方法用于启动 Activity，执行后，返回到启动它的 Activity 来执行回调函数，代码如下所示：

```
package com.redarmy.extra;


import android.app.Activity;
import android.content.Intent;
import android.os.Bundle;
import android.view.View;
import android.view.View.OnClickListener;
import android.widget.Button;


public class ActivityMain extends Activity implements OnClickListener{
  public Button button1;
  public Button button2;
  public static final int REQUEST_CODE=1;
   @Override
   public void onCreate(Bundle savedInstanceState) {
      super.onCreate(savedInstanceState);
      setContentView(R.layout.main);
```

```
        setTitle("这是在 ActivityMain");
        button1 = (Button)findViewById(R.id.Button01);
        button2 = (Button)findViewById(R.id.Button02);
        //添加监听
        button1.setOnClickListener(this);
        button2.setOnClickListener(this);
    }
    @Override
    public void onClick(View v) {
        int temp =v.getId();
        switch (temp) {
        case R.id.Button01:
            Intent intent1 = new Intent(ActivityMain.this,
ActivityOther. class);
            intent1.putExtra("activityMain", "数据来自 activityMain");
            startActivityForResult(intent1, REQUEST_CODE);
            break;
        case R.id.Button02:
            break;
        }
    }
     @Override
    protected void onActivityResult(int requestCode, int resultCode,
Intent data) {
            if (requestCode == REQUEST_CODE) {
                if (resultCode == RESULT_CANCELED)
                    setTitle("取消");
                else if (resultCode == RESULT_OK) {
                    String temp=null;
                     Bundle extras = data.getExtras();
                        if (extras != null) {
                        temp = extras.getString("ActivityOther");
                        }
                    setTitle(temp);
                }
            }
        }
}
```

ActivityOther 返回值的代码如下所示：

```
package com.redarmy.extra;

import android.app.Activity;
```

```
import android.content.Intent;
import android.os.Bundle;
import android.view.View;
import android.view.View.OnClickListener;
import android.widget.Button;

public class ActivityOther extends Activity {

    @Override
    protected void onCreate(Bundle savedInstanceState) {
        // TODO Auto-generated method stub
        super.onCreate(savedInstanceState);
        setContentView(R.layout.otheractivity);
        Button button3 = (Button)findViewById(R.id.Button03);
        button3.setOnClickListener(new OnClickListener() {
            @Override
            public void onClick(View v) {
                Bundle bundle = new Bundle();
                bundle.putString("ActivityOther","数据来自ActivityOther");
                Intent mIntent = new Intent();
                mIntent.putExtras(bundle);
                setResult(RESULT_OK, mIntent);
                finish();
            }
        });
        String data=null;
        Bundle extras = getIntent().getExtras();
            if (extras != null) {
                data = extras.getString("activityMain");
            }
        setTitle("现在是在ActivityOther里:"+data);
    }
}
```

上述代码展示了数据传递的过程，请读者多加练习。

5.2 Intent 详解

在 Android 参考文档中，对 API 的定义是：执行某操作的一个抽象描述（确实很抽象），其基本内容分为 Action、Data、Type、Category、Extra 和 Component 6 个部分，分述如下。

（1）对执行操作的描述：操作（Action）。

（2）对于本次动作相关的数据进行描述：数据（Data）。

（3）对数据类型的描述：数据类型（Type）。

（4）对执行动作的附加信息进行描述：类别（Category）。

（5）对执行动作的附加信息的描述：附件信息（Extras）。

对目标组件的描述：目标组件(Component)。

5.2.1　操作（Action）

Action 描述 Intent 触发动作名字的字符串。对于 BroadcastIntent 来说，Action 指被广播出去的动作。理论上，Action 可以为任何字符串；与 Android 系统应用有关的 Action 字符串以静态字符串常量的形式定义在 Intent 类中。表 5-1 列出了当前 Android 系统中常见的 Activity Action Intent 的 Action。表 5-2 列出了常见的 BroadcastIntent Action 常量。

表 5-1

Activity Action Intent 字符串常量	描　述
ACTION_CALL	拨出 Data 里指定的电话号码
ACTION_EDIT	打开编辑 Data 里指定数据对应的应用程序
ACTION_MAIN	主程序入口，不会接收数据，结束后不返回数据
ACTION_SYNC	在 Android 平台和服务器之间同步数据
ACTION_VIEW	根据 Data 的不同类型，打开对应的应用程序显示数据
ACTION_DIAL	启动系统拨号程序或其他拨号程序，并显示 Data 里指定的电话号码
ACTION_SENDTO	向 Data 描述的目标地址发送数据

表 5-2

BroadcastIntent Action 字符串常量	描　述
ACTION_TIME_TICK	系统时间每过 1 分钟发出的广播
ACTION_TIME_CHANGED	系统时间通过设置发生了变化
ACTION_TIMEZONE_CHANGED	时区改变
ACTION_BOOT_COMPLETED	系统启动完毕
ACTION_PACKAGE_ADDED	新的应用程序 APK 包安装完毕
ACTION_PACKAGE_CHANGED	现有应用程序 APK 包改变
ACTION_PACKAGE_REMOVED	现有应用程序 APK 包被删除
ACTION_UID_REMOVED	用户 ID 被删除

5.2.2　数据（Data）（与动作相关联的数据）

Data 描述 Intent 要操作的数据 URI 和数据类型。有的动作需要对相应的数据进行处理。比如，对于动作 ACTION_EDIT 来说，其数据可以是联系人、短信息等可编辑 URI；而对于 ACTION_CALL 来说，其数据可以是 tel://格式的电话号码 URI。

正确设置 Intent 的数据，对于 Android 寻找系统中匹配 Intent 请求的组件很重要。如果使用 ACTION_CALL，但其数据设置成**mailto:**//格式的 URI，那么所期望的“启动打电话应用程序”这一动作会因为没有相对应的应用程序而不被执行。所以，每次使用 Intent 时，都应该留意与设置的 Action 相关的数据类型和格式。

这种 URI 表示通过 ContentURI 类描述，具体内容请参考 android.net.ContentURI 类的文档。以联系人为例，以下是一些 action/data 对及其要表达的意图：

（1）VIEW_ACTION content://contacts/1，显示标识符为“1”的联系人的详细信息。

（2）EDIT_ACTION content://contacts/1，编辑标识符为“1”的联系人的详细信息。

（3）VIEW_ACTION content://contacts/，显示所有联系人列表。

（4）PICK_ACTION content://contacts/，显示所有联系人列表，并且允许用户在列表中选择一个，然后将其返回给父 Activity。例如，电子邮件客户端使用该 Intent，要求用户在联系人列表中选择一个。

5.2.3 类型（Type）

数据类型（Type）显示指定 Intent 的数据类型（MIME）。一般情况下，Intent 的数据类型可根据数据本身来判定，通过设置该属性，可以强制采用显示指定的类型，而不再推导。

5.2.4 类别（Category）

类别（Category）是对被请求组件的额外描述信息。Android 在 Intent 类中定义了一组静态字符串常量表示 Intent 的不同类别。表 5-3 列出了常见的 Category 常量。

表 5-3

Category 字符串常量	描 述
CATEGORY_BROWSABLE	目标 Activity 可通过在网页浏览器中单击链接而激活(比如，单击浏览器中的图片链接)
CATEGORY_GADGET	表示目标 Activity 可以被嵌入到其他 Activity 当中
CATEGORY_HOME	目标 Activity 是 HOME Activity，即手机开机启动后显示的 Activity，或按下 HOME 键后显示的 Activity
CATEGORY_LAUNCHER	表示目标 Activity 是应用程序中最优先被执行的 Activity
CATEGORY_PREFERENCE	表示目标 Activity 是一个偏好设置的 Activity

5.2.5 附件信息（Extras）

使用 Intent 连接不同的组件时，有时需要在 Intent 中附加额外的信息，以便将数据传递给目标 Activity。比如，ACTION_TIMEZONE_CHANGED 需要带有附加信息表示新的时区。Extra 用键值对结构保存在 Intent 对象中，Intent 对象通过调用方法 putExtras()和 getExtas()来存储和获取 Extra。Extra 以 Bundle 对象的形式保存。Bundle 对象提供了一系列 put 和 get 方法来设置、提取相应的键值信息。在 Intent 类中，同样为 Android 系统应用的一些 Extra 键值定义了静态字符串常量。

表 5-4

Extra 键值字符串常量	描 述
EXTRA_BCC	装有邮件密送地址的字符串数组
EXTRA_CC	装有邮件抄送地址的字符串数组
EXTRA_EMAL	装有邮件发送地址的字符串数组
EXTRA_INTENT	使用 ACTION_PICK_ACTIVITY 动作时，装有 Intent 选项的键
EXTRA_KEY_EVENT	触发该 Intent，按键的 KeyEvent 对象
EXTRA_PHONE_NUMBER	使用拨打电话相关 Action 时，电话号码字符串的键，类型为 String
EXTRA_SHORTCUT_ICON EXTRA_SHORTCUT_ICON_RESOURCE EXTRA_SHORTCUT_INTENT EXTRA_SHORTCUT_NAME	使用 ACTION_CREATE_SHORTCUT 在 HomeActivity 创建快捷方式时，对快捷方式的描述信息。其中，ICON 和 ICON_RESOURCE 描述的是快捷方式的图标，类型分别为 Bitmap 和 ShortcutIconResource。INTENT 描述的是跨界方式对应的 Intent 对象。NAME 描述的是快捷方式的名字
EXTTRA_SUBJECT	描述信息主题的键
EXTRA_TEXT	使用 ACTION_SEND 动作时，用来描述要发送的文本信息，类型为 CharSequence
EXTRA_TITLE	使用 ACTION_CHOOSER 动作时，描述对话框标题的键，类型为 CharSequence
EXTRA_UID	使用 ACTION_UID_REMOVED 动作时，描述删除的用户 ID 的键，类型为 int

5.2.6　目标组件（Component）

组件名称是指 Intent 目标组件的名称。组件名称是 ComponentName 对象，这种对象名称是目标组件类名和目标组件所在应用程序的包名的组合。组件包名不一定要和 manifest 文件中的包名完全匹配。组件名称是一个可选项。如果 Itent 消息中指明了目标组件的名称，这就是一个显式消息，Intent 会传递给指明的组件。如果目标组件名称没有指定，Android 通过 Intent 内的其他信息和已注册的 IntentFilter 的比较来选择合适的目标组件。

总之，Action、Data/Type、Category 和 Extras 一起形成了一种语言规范，使系统能够理解诸如"查看某联系人的详细信息"或"给某人打电话"之类的短语。随着应用不断加入到系统中，Android 系统可以添加新的 Action、Data/Type、Category 来扩展语言。当然，最受益的还是应用本身，可以利用这种语言机制来处理不同的动作和数据。

5.3　解析 Intent

Intent 是一种在不同组件之间传递的请求消息，是应用程序发出的请求和意图。对于一个完整的消息机制，Intent 不仅需要发送端，还需要接收端。Android 如何解析 Intent 的请求内容，并选择合适的组件响应 Intent 请求呢?下面将深入地分析。

5.3.1　显式 Intent 与隐式 Intent

明确指出了目标组件名称的 Intent，称为显式 Intent。没有明确指出目标组件名称的 Intent，称为隐式 Intent。Android 系统使用 IntentFilter 来寻找与隐式 Intent 相关的对象。

显式 Intent 直接用组件的名称定义目标组件。这种方式很直接，但是由于开发人员往往并不清楚其他应用程序的组件名称，因此显式 Intent 更多用于在应用程序内部传递消息。比如在某应用程序内，一个 Activity 启动一个 Service。隐式 Intent 恰恰相反，它不会用组件名称定义需要激活的目标组件，它更广泛地用于在不同应用程序之间传递消息。

清楚了显式 Intent 和隐式 Intent 的概念后，下面讨论决定 Intent 目标组件的因素。在显式 Intent 消息中，决定目标组件的唯一要素就是组件名称，因此，如果 Intent 中明确定义了目标组件的名称，完全不用再定义其他 Intent 内容。对于隐式 Intent 则不同，由于没有明确的目标组件名称，所以必须由 Android 系统帮助应用程序寻找与 Intent 请求意图匹配的组件，方法是：Android 将 Intent 的请求内容和一个叫做 IntentFilter 的过滤器相比较，IntentFilter 中包含系统中的所有可能的待选组件。如果 IntentFilter 中的某一组件匹配隐式 Intent 请求的内容，Android 就选择它作为隐式 Intent 的目标组件。

Android 如何知道应用程序能够处理某种类型的 Intent 请求呢？这需要应用程序在 AndroidManifest.xml 中声明所含组件的过滤器（即可以匹配哪些 Intent 请求）。没有声明 IntentFilter 的组件只能响应指明自己名字的显式 Intent 请求，而无法响应隐式 Intent 请求。声明了 IntentFilter 的组件既可以响应显式 Intent 请求，也可以响应隐式 Intent 请求。在通过和 IntentFilter 比较来解析隐式 Intent 请求时，Android 将以下三个因素作为选择的参考标准：Action、Data 和 Category。Extra 和 Flag 在解析收到 Intent 时不起作用。

5.3.2　IntentFilter

为了告诉 Android 自己能响应、处理哪些隐式 Intent 请求，应用程序的组件可以声明一个甚至多个 IntentFilter。每个 IntentFilter 描述该组件所能响应 Intent 请求的能力——组件希望接收什么类型的请求行为，什么类型的请求数据。比如在请求网页浏览器的例子中，网页浏览器程序的 IntentFilter 声明希望接收的 Intent Action 是 WEB_SEARCH_ACTION，与之相关的请求数据是网页地址 URI 格式。

如何为组件声明自己的 IntentFilter？常见的方法是在 AndroidManifest.xml 文件中用属性<Intent-Filter>描述组件的 IntentFilter。

如前所述，隐式 Intent 和 IntentFilter 进行比较的三要素是 Intent 的 Action、Data 以及 Category。实际上，隐式 Intent 请求要能够传递给目标组件，必要通过这三个方面的检查。如果任何一个方面不匹配，Android 都不会将该隐式 Intent 传递给目标组件。下面介绍这三方面检查的规则。

1．动作测试

<intent-filter>元素中可以包括子元素<action>，比如：

```
<intent-filter>
    <action android:name=”com.example.project.SHOW_CURRENT” />
    <action android:name=”com.example.project.SHOW_RECENT” />
    <action android:name=”com.example.project.SHOW_PENDING” />
</intent-filter>
```

一个<intent-filter>元素至少应该包含一个<action>，否则，任何 Intent 请求都不能和该<intent-filter>匹配。

如果 Intent 请求的 Action 和<intent-filter>中的某一条<action>匹配，那么该 Intent 就通过了这条<intent-filter>的动作测试。

如果 Intent 请求或<intent-filter>中没有说明具体的 Action 类型，会出现下面两种情况。

（1）如果<intent-fitler>中没有包含任何 Action 类型，那么无论什么 Intent 请求都无法和这条<intent-filter>匹配。

（2）反之，如果 Intent 请求中没有设定 Action 类型，只要<intent-filter>中包含 Action 类型，这个 Intent 请求就将顺利地通过<intent-filter>的行为测试。

2．类型测试

<intent-filter>元素可以包含<category>子元素，比如：

```
<intent-filter………>
    <category android:name=”android.Intent.Category.DEFAULT” />
    <category android:name=”android.Intent.Category.BROWSABLE” />
</intent-filter>
```

只有当 Intent 请求中所有的 Category 与组件某一个 IntentFilter 的<category>完全匹配时，才会让该 Intent 请求通过测试。IntentFilter 中多余的<category>声明不会导致匹配失败。一个没有指定任何类别测试的 IntentFilter 只会匹配没有设置类别的 Intent 请求。

3．数据测试

数据在<intent-filter>中的描述如下：

```
<intent-filter………>
    <data android:type=”video/mpeg” android:scheme=”http” ……/>
    <data android:type=”audio/mpeg” android:scheme=”http” ……/>
</intent-filter>
```

<data>元素指定了希望接收的 Intent 请求的数据 URI 和数据类型。URI 被分成 3 个部分来匹配：scheme、athority 和 path。其中，用 setData()设定的 Intent 请求的 URI 数据类型和 scheme 必须与 IntentFilter 中指定的一致。若 IntentFilter 中还指定了 authority 或 path，它们也要匹配

才能通过测试。

下面用 Intent 激活 Android 自带的电话拨号程序，以此实例说明，使用 Intent 不像其概念描述得那样难。

第一步：新建 Android 项目，名为 PhoneDemo。

第二步：修改 main.xml 文件，代码如下所示。

```
<?xml version="1.0" encoding="utf-8"?>
<LinearLayout
xmlns:android="http://schemas.android.com/apk/res/android"
   android:orientation="vertical"
   android:layout_width="fill_parent"
   android:layout_height="fill_parent"
   >
<Button android:id="@+id/Button01"
android:layout_height="wrap_content"
android:layout_width="fill_parent"
android:layout_marginLeft="20dip"
android:layout_marginRight="20dip"
android:text="@string/phone"></Button>
</LinearLayout>
```

第三步：主类代码如下所示：

```
package com.redarmy.phone;

import android.app.Activity;
import android.content.Intent;
import android.net.Uri;
import android.os.Bundle;
import android.view.View;
import android.view.View.OnClickListener;
import android.widget.Button;

public class ActivityMain extends Activity {
   /** Called when the activity is first created. */
   @Override
   public void onCreate(Bundle savedInstanceState) {
      super.onCreate(savedInstanceState);
      setContentView(R.layout.main);
      Button button =(Button)findViewById(R.id.Button01);
      button.setOnClickListener(new OnClickListener() {
         @Override
         public void onClick(View v) {
            Intent intent = new
Intent(Intent.ACTION_DIAL,Uri.parse ("tel://13800138000"));
            startActivity(intent);
```

```
        }
    });
  }
}
```

以上列出的是 Android 系统自带拨号程序的 Intent。系统自带程序的 Intent 很多，希望读者多加练习，并且查阅相关文档及 API 来巩固所学知识。

5.4 Android 中的广播机制

Intent 可以用来启动一个新的 Activity，但是 Intent 的作用远不止于此，它还有一个重要的机制，就是作为不同进程间传递数据和事件的媒体。

通常用户自己的应用或者 Android 系统本身在某些事件来临时，会将 Intent 广播出去，注册的 BroadcastReceiver 可以监听到这些 Intent，并且获得保存在 Intent 中的数据。

BroadcastReceiver 是用户接收广播通知的组件。广播是一种同时通知多个对象的事件通知机制，Android 中的广播通知要么来自系统，要么来自普通应用程序。很多事件可能导致系统广播，比如手机所在的时区发生变化，电池电量低，用户改变系统语言设置等。当然，也有广播来自应用程序，比如一个应用程序通知其他应用程序某些数据下载完毕。

为了响应不同事件的通知，应用程序可以注册不同的 BroadcastReceiver。所有的 BroadcastReceiver 都继承自基类 BroadcastReceiver。需要说明的是，BroadcastReceiver 自身并不实现图形用户界面，但是当它收到某个通知消息后，可以启动 Activity 作为响应，或者通过 NotificationManager 提醒用户。

在 Android 平台中，Broadcast 是一种广泛运用的在应用程序之间传输信息的机制。BroadcastReceiver 是对发送的 Broadcast 进行过滤接收并响应的一类组件。下面详细介绍发送 Broadcast 和使用 BroadcastReceiver 过滤接收的过程。

首先，在需要发送信息的地方把要发送的信息和用于过滤的信息（如 Action、Category）装入一个 Intent 对象；然后，通过调用 Context.sendBroadcast()、sendOrderBroadcast()或 sendStickyBroadcast()方法，把 Intent 对象以广播的形式发送出去。当 Intent 发送以后，所有已经注册的 BroadcastReceiver 检查注册时的 IntentFileter 是否与发送的 Intent 相匹配，若匹配，将会调用 BroadcastReceiver 的 onReceive()方法。所以，定义一个 BroadcastReceiver 的时候，需要自己实现 onReceive()方法。

通常使用 sendBroadcast()或 sendStickyBroadcast()发送出去的 Intent，所有满足条件的 BroadcastReceiver 都会随机地执行其 onReceive()方法。通过 sendOrderBroadcast()方法发送出去的 Intent 会根据 BroadcastReceiver 注册时 IntentFilter 设置的优先级顺序来执行，相同优先级的 BroadcastReceiver 随机执行。对于 sendStickyBroadcast()，Intent 在发送后一直存在，并且在以后调用 registerReceiver()注册匹配的 Intent 时会把该 Intent 直接返回。

注册 BroadcastReceiver 有下述两种方式。

（1）静态地在 AndroidManifest.xml 中用<receiver>标签声明注册，并在标签内部用<intent-filter>标签设置过滤器。推荐采用这种方法，这也是最常用的方法。代码如下所示：

```
<receiver android:name="MyReceiver">
<intent-filter>
<action android:name="com.redarmy.action.NEW_BROADCAST"></action>
```

```
    </intent-filter>
</receiver>
```

（2）动态地在代码中先定义并设置好一个 IntentFilter 对象，然后在需要注册的地方调用 Context.registerReceiver()方法。代码如下所示：

```
IntentFilter filter = new IntentFilter(NEW_BROADCAST);
MyReceiver  myReceiver = new MyReceiver();
registerReceiver(myReceiver, filter);
```

如果想将一个已经注册的 BroadcastReceiver 注销，采用下述代码：

```
unregisterReceiver(myReceiver);
```

如果将用动态方式注册的 BroadcastReceiver 的 Context 对象销毁，BroadcastsReceiver 将自动取消注册。

另外，若在使用 sendBroadcast()方法时指定了接收的权限，只有在 AndroidManifest.xml 中用<uses-permission>标签声明了拥有此权限的 BroadcastReceiver，才有可能接收到发送来的 Broadcast。同样地，若注册 BroadcastReceiver 时指定了可接收 Broadcast 的权限，只有在包内的 AndroidMainfest.xml 中用<uses-permission>标签声明了拥有此权限的 Context 对象所发送的 Broadcast，才可能被该 BroadcastReceiver 接收。

Android 系统会在一些系统状态发生变化时，对变化内容用 Broadcast 的形式发送给其他应用程序，以便执行自定义操作。当注册响应系统变化的 BroadcastReceiver 时，需要设置相应的 IntentFilter，以匹配系统发送 Intent。请参考 Andorid API，其中列出了常见的系统状态发生变化时发送 Broadcast 的 Intent。

5.5　Intent 实现广播案例

本节介绍，一个广播事件的案例。

第一步：新建 Android 项目，名称为 MyReceiver，并在 ActivityMain.java 中输入以下代码：

```
package com.redarmy.receiver;


import android.app.Activity;
import android.content.Intent;
import android.content.IntentFilter;
import android.os.Bundle;
import android.view.Menu;
import android.view.MenuItem;

public class ActivityMain extends Activity {

   private static final int MENU_SAVE = Menu.FIRST;
   private static final int MENU_DELETE = Menu.FIRST + 1;

   public static final String MYACTION =
```

```
"com.3gdci.action.NEW_ BROADCAST_ONE";
    public static final String MYACTION1 =
"com.3gdci.action.NEW_ BROADCAST_TWO";

    @Override
    public void onCreate(Bundle savedInstanceState) {
        super.onCreate(savedInstanceState);
        setContentView(R.layout.main);

    }
    @Override
    public boolean onCreateOptionsMenu(Menu menu) {
        super.onCreateOptionsMenu(menu);
        menu.add(0, MENU_SAVE, 0, "保存")
                .setIcon(android.R.drawable.ic_menu_save);
        menu.add(0, MENU_DELETE, 0, "删除").setIcon(
                android.R.drawable.ic_menu_delete);
        return true;
    }

    @Override
    public boolean onOptionsItemSelected(MenuItem item) {
        // 得到当前选中的MenuItem的ID
        int item_id = item.getItemId();
        switch (item_id) {
        case MENU_SAVE:
            Intent intent1 = new Intent(MYACTION);
            sendBroadcast(intent1);
            ActivityMain.this.finish();
            break;
        case MENU_DELETE:
            Intent intent2 = new Intent(MYACTION1);
            sendBroadcast(intent2);
            ActivityMain.this.finish();

            break;
        }
        return true;
    }

}
```

第二步：新建 MyReceiver 与 MyReceiver1，并且都继承了 BroadcastReceiver，重写

onReceive(Context context, Intent intent)方法。

第一个 Java 类的代码如下所示：

```
package com.redarmy.receiver;
import android.app.Notification;
import android.app.NotificationManager;
import android.app.PendingIntent;
import android.content.BroadcastReceiver;
import android.content.Context;
import android.content.Intent;
public class MyReceiver extends BroadcastReceiver {
   public Context context;
   public static final int NOTID = 123456;
   @Override
   public void onReceive(Context context, Intent intent) {
      this.context = context;
      showNotificationMessage();
   }
   private void showNotificationMessage() {
      // 第一步：得到 NotificationManager 的一个引用
      NotificationManager NotManager = (NotificationManager) context
   .getSystemService(android.content.Context.NOTIFICATION_SERVICE);
      //第二步：初始化一个 Notification
      long when = System.currentTimeMillis();//得到系统当前时间
      //构造器中参数的含义是：1.显示的图片 id;2.显示的文本;3.时间
      Notification notification = new Notification(R.drawable.icon,"
在 MyReceiver 中",when);
      //第三步:设置 Notification 的参数
      Intent notIntent = new Intent(context,ActivityMain.class);
      PendingIntent pIntent = PendingIntent.getActivity(context, 0,
notIntent, 0);
      notification.setLatestEventInfo(context, "在 MyReceiver 中",
null, pIntent);
      //第四步：显示
      NotManager.notify(NOTID, notification);
   }
}
```

第二个 Java 类的代码如下所示：

```
package com.redarmy.receiver;

import android.app.NotificationManager;
import android.content.BroadcastReceiver;
import android.content.Context;
import android.content.Intent;
public class MyReceiver1 extends BroadcastReceiver {
```

```
    Context context;
    @Override
    public void onReceive(Context context, Intent intent) {
        // TODO Auto-generated method stub
        this.context = context;
        DeleteNotification();
    }
    private void DeleteNotification() {
        NotificationManager notificationManager= (NotificationManager)
context
    .getSystemService(android.content.Context.NOTIFICATION_SERVICE);
        notificationManager.cancel(MyReceiver.NOTID);
    }
}
```

第三步：在 AndroidManifest.xml 文件中添加注册，代码如下所示：

```
<?xml version="1.0" encoding="utf-8"?>
<manifest xmlns:android="http://schemas.android.com/apk/res/android"
    package="com.redarmy.receiver" android:versionCode="1"
    android:versionName="1.0">
    <application android:icon="@drawable/icon"
android:label= "@string/ app_name">
        <activity android:name=".ActivityMain"
android:label= "@string/ app_name">
            <intent-filter>
                <action android:name="android.intent.action.MAIN" />
                <category android:name="android.intent.category.LAUNCHER"/>
            </intent-filter>
        </activity>
    <receiver android:name="MyReceiver">
        <intent-filter>
    <action android:name="com.3gdci.action.NEW_BROADCAST_ONE"></action>
        </intent-filter>
</receiver>
<receiver android:name="MyReceiver1">
<intent-filter>
  <action android:name="com.3gdci.action.NEW_BROADCAST_TWO"></action>
  </intent-filter>
</receiver>
</application>
    <uses-sdk android:minSdkVersion="3" />
</manifest>
```

第 6 章　Android 的数据存储操作

数据存储是应用程序最基本的问题，任何企业系统、应用软件都必须解决。数据必须以某种方式保存，不能丢失，并且要能够被有效、简便地使用和更新。本章将详细介绍 Android 中的数据存储方式以及数据共享问题。

6.1　Android 数据存储概述

Android 系统提供了 4 种数据存储方式，但是由于存储的数据都是应用程序私有的，所以如果需要在其他应用程序中使用这些数据，应采用 Android 提供的 Content Providers(数据共享)。Android 中的 4 种数据存储方式分述如下：

（1）Shared Preferences：用来存储 Key-value paires 格式的数据。它是一个轻量级的键值存储机制，只可以存储基本数据类型的数据。

Shared Preferences 主要针对系统配置信息的保存。例如，给程序界面设置了音效，想在下一次启动时，保留上次设置的音效。由于 Android 采用 Linux 核心，所以 Android 系统的界面采用 Activity 栈的形式，在系统资源不足时会收回一些界面，因此有些操作需要在不活动时保留下来，以便再次激活时、能够显示出来。

（2）Files：通过 FileInputStream 和 FileOutputStream 对文件进行操作。但是在 Android 中，文件是应用程序私有的，一个应用程序无法读写其他应用程序的文件。

Files 把需要保存的文件通过文件的形式记录下来，当需要这些数据时，通过读取该文件获得数据。因为 Android 采用 Linux 核心，所以系统中的文件也是 Linux 形式。

（3）SQLite:Android 提供了一个标准的数据库，支持 SQL 语句。

SQLite 是一个开源的关系型数据库，与普通关系数据库一样，也具有 ACID 的特性。它可以用来存储大量的数据，并且能够很容易地对数据执行使用、更新、维护等操作，但是操作规范前两种方式复杂。

（4）NetWork：通过网络存储和获得数据。

Network 用于将数据存储于网络，还需要使用 java.net.*和 android.net.*这些类。后面将通过实例详细介绍。

前面的章节提到过 Content Providers（数据共享），因为在 Android 中，很多数据不只是供一个应用程序使用，可能要被多个应用程序共同使用。由于 Shared Preference 存储的是与应用程序系统配置相关的数据，只供本应用程序使用，要想共享数据，需要采用 Files、SQLite 以及网络存储等方式来存储数据。

了解不同的数据存储方式之后，就可以根据应用程序的需要来选择。下面分别介绍这些存储方式。

6.2　Shared Preferences 存储

Shared Preferences 类似于常用的 ini 文件，用来保存应用程序的一些属性设置，在 Android 平台常用于存储较简单的参数设置。例如，可以通过它保存上一次用户所做的修改或者自定义参数设定，当再次启动程序后，依然保持原有的设置。通过 getPreferences()方法获得 Preferences 对象，通过 SharedPreferences.Editor editor = uiState.edit()取得编辑对象，然后通过 editor.put...()方法添加数据，最后通过 commit()方法保存数据。如果不需要与其他模块共享数据，使用 Activity.getPreferences()方法保持数据的私有。需要强调的是，无法直接在多个程序间共享 Preferences 数据。

下面通过一个实例来探讨 Shared Preferences。首先，运行应用程序，如图 6-1 所示。

图 6-1

当前音效处于“关闭”状态时，按“上”方向键开启音效，然后退出程序。再次启动应用程序，当前音效处于“开”的状态，如图 6-2 所示。

图 6-2

在此过程中，退出应用程序时，音效状态保存在 Preferences 中，因此音效一启动就读取出上次的数据，开始播放音乐。代码如下所示：

```
package com.redarmy.sharedPre;

import android.app.Activity;
import android.content.SharedPreferences;
import android.os.Bundle;
import android.view.KeyEvent;
import android.widget.TextView;

public class ActivityMain extends Activity
{
```

```
    private MIDIPlayer  mMIDIPlayer  = null;
    private boolean     mbMusic      = false;
    private TextView mTextView = null;

    /** Called when the activity is first created. */
    @Override
    public void onCreate(Bundle savedInstanceState)
    {
        super.onCreate(savedInstanceState);
        setContentView(R.layout.main);

        mTextView = (TextView) this.findViewById(R.id.TextView01);

        mMIDIPlayer = new MIDIPlayer(this);

        /* 装载数据 */
        // 取得活动的 preferences 对象
        SharedPreferences settings=getPreferences(Activity.MODE_PRIVATE);

        // 取得值.
        mbMusic = settings.getBoolean("bmusic", false);

        if (mbMusic)
        {
            mTextView.setText("当前音乐状态：开");
            mbMusic = true;
            mMIDIPlayer.PlayMusic();
        }
        else
        {
            mTextView.setText("当前音乐状态：关");
        }

    }


    public boolean onKeyUp(int keyCode, KeyEvent event)
    {
        switch (keyCode)
        {
            case KeyEvent.KEYCODE_DPAD_UP:
```

```
            mTextView.setText("当前音乐状态：开");
            mbMusic = true;
            mMIDIPlayer.PlayMusic();
            break;
        case KeyEvent.KEYCODE_DPAD_DOWN:
            mTextView.setText("当前音乐状态：关");
            mbMusic = false;
            mMIDIPlayer.FreeMusic();
            break;
    }
    return true;
}


public boolean onKeyDown(int keyCode, KeyEvent event)
{
    if (keyCode == KeyEvent.KEYCODE_BACK)
    {
        /* 退出应用程序时保存数据 */
        // 取得活动的 preferences 对象.
        SharedPreferences uiState = getPreferences(0);

        // 取得编辑对象
        SharedPreferences.Editor editor = uiState.edit();

        // 添加值
        editor.putBoolean("bmusic", mbMusic);

        // 提交保存
        editor.commit();

        if ( mbMusic )
        {
            mMIDIPlayer.FreeMusic();
        }
        this.finish();
        return true;
    }
    return super.onKeyDown(keyCode, event);
 }
}
```

类代码如下所示：

```
package com.redarmy.sharedPre;

import java.io.IOException;

import android.content.Context;
import android.media.MediaPlayer;

public class MIDIPlayer
{
   public MediaPlayer  playerMusic   = null;
   private Context     mContext      = null;


   public MIDIPlayer(Context context)
   {
      mContext = context;
   }

   /* 播放音乐 */
   public void PlayMusic()
   {
      /* 装载资源中的音乐 */
      playerMusic = MediaPlayer.create(mContext, R.raw.start);

      /* 设置是否循环 */
      playerMusic.setLooping(true);
      try
      {
         playerMusic.prepare();
      }
      catch (IllegalStateException e)
      {
         e.printStackTrace();
      }
      catch (IOException e)
      {
         e.printStackTrace();
      }
      playerMusic.start();
   }

   /* 停止并释放音乐 */
```

```
    public void FreeMusic()
    {
        if (playerMusic != null)
        {
            playerMusic.stop();
            playerMusic.release();
        }
    }
}
```

6.3 Files 存储

在 Android 系统中，可以在设备本身的存储器或者外接的存储器中创建用于保存数据的文件。同样地，在默认状态下，文件不能在不同的程序间共享。用文件存储数据，可以通过 openFileOutput 方法打开一个文件（如果文件不存在，将自动创建），通过 load 方法获取文件中的数据。通过 deleteFile 方法可以删除指定的文件。

下面用文件的方式来保存 6.2 节中介绍的音乐状态。首先，在退出应用程序之前，通过自己定义的 save 方法保存数据（音效状态）；然后，在应用程序启动时，通过 load 方法读取数据。代码如下所示：

```
package com.redarmy.file;

import java.io.FileInputStream;
import java.io.FileNotFoundException;
import java.io.FileOutputStream;
import java.io.IOException;
import java.util.Properties;

import android.app.Activity;
import android.content.Context;
import android.os.Bundle;
import android.view.KeyEvent;
import android.widget.TextView;

public class ActivityMain extends Activity
{

    private MIDIPlayer  mMIDIPlayer  = null;
    private boolean     mbMusic      = false;
    private TextView mTextView = null;
    /** Called when the activity is first created. */
    @Override
    public void onCreate(Bundle savedInstanceState)
```

```
{
    super.onCreate(savedInstanceState);
    setContentView(R.layout.main);
    mTextView = (TextView) this.findViewById(R.id.TextView01);
    mMIDIPlayer = new MIDIPlayer(this);
    /* 读取文件数据  */
    load();
    if (mbMusic)
    {
        mTextView.setText("当前音乐状态: 开");
        mbMusic = true;
        mMIDIPlayer.PlayMusic();
    }
    else
    {
        mTextView.setText("当前音乐状态: 关");
    }
}
public boolean onKeyUp(int keyCode, KeyEvent event)
{
    switch (keyCode)
    {
        case KeyEvent.KEYCODE_DPAD_UP:
            mTextView.setText("当前音乐状态: 开");
            mbMusic = true;
            mMIDIPlayer.PlayMusic();
            break;
        case KeyEvent.KEYCODE_DPAD_DOWN:
            mTextView.setText("当前音乐状态: 关");
            mbMusic = false;
            mMIDIPlayer.FreeMusic();
            break;
    }
    return true;
}

public boolean onKeyDown(int keyCode, KeyEvent event)
{
    if (keyCode == KeyEvent.KEYCODE_BACK)
    {
        /* 退出应用程序时保存数据 */
        save();
```

```
            if ( mbMusic )
            {
                mMIDIPlayer.FreeMusic();
            }
            this.finish();
            return true;
        }
        return super.onKeyDown(keyCode, event);
    }

    /* 装载、读取数据 */
    void load()
    {
        /* 构建 Properties 对象 */
        Properties properties = new Properties();
        try
        {
            /* 开发文件 */
            FileInputStream stream = this.openFileInput("music.cfg");
            /* 读取文件内容 */
            properties.load(stream);
        }
        catch (FileNotFoundException e)
        {
            return;
        }
        catch (IOException e)
        {
            return;
        }
        /* 取得数据 */
        mbMusic = Boolean.valueOf(properties.get("bmusic").toString());
    }

    /* 保存数据 */
    boolean save()
    {
        Properties properties = new Properties();

        /* 将数据打包成 Properties */
        properties.put("bmusic", String.valueOf(mbMusic));
```

```
        try
        {
            FileOutputStream stream = this.openFileOutput("music.cfg",
Context.MODE_WORLD_WRITEABLE);

            /* 将打包的数据写入文件*/
            properties.store(stream, "");
        }
        catch (FileNotFoundException e)
        {
            return false;
        }
        catch (IOException e)
        {
            return false;
        }
        return true;
    }
}
```

6.4　Network 存储

通过网络来获取和保存数据资源，要求设备保持网络连接状态，所以有一些限制。将数据存储到网络上的方法很多，比如将数据以文件的方式上传到服务器、发送邮件等。本节将在应用程序退出时，把数据发送到电子邮箱中备份，在模拟器中配置电子邮件账户后将邮件发送出去。配置电子邮件账户的操作如下所述。

（1）启动模拟器，打开“菜单”，选择“电子邮件”项，启动之后，填写邮件地址和密码，然后选中“默认情况下从该账户发送电子邮件”，如图 6-3 所示。

图　6-3

（2）单击“下一步”按钮，程序将自动配置电子邮件的相关信息，如图 6-4 所示。

（3）自动配置完成后，设置名称等信息，如图 6-5 所示，电子邮箱配置成功。

图　6-4

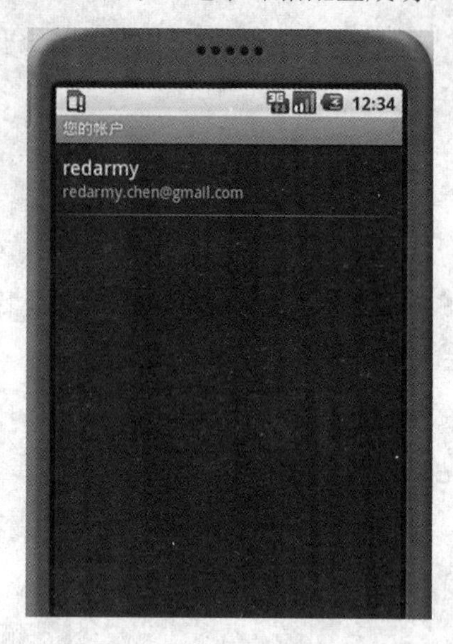

图　6-5

配置电子邮件账户之后，可以通过程序来调用模拟器的电子邮件客户端发送邮件了。在 Android 系统中发送电子邮件，是通过 startActivity 方法来调用要发送的邮件数据的 Intent。采用 putExtra 方法设置邮件的主题、内容、附件等，代码如下所示：

```
package com.redarmy.network;

import android.app.Activity;
import android.content.Intent;
import android.net.Uri;
import android.os.Bundle;
import android.view.KeyEvent;

public class ActivityMain extends Activity
{
   private int  miCount = 0;
   /** Called when the activity is first created. */
   @Override
   public void onCreate(Bundle savedInstanceState)
   {
      super.onCreate(savedInstanceState);
      setContentView(R.layout.main);

      miCount=1000;
   }
```

```
    public boolean onKeyDown(int keyCode, KeyEvent event)
    {
        if (keyCode == KeyEvent.KEYCODE_BACK)
        {
            //退出应用程序时保存数据
            /* 发送邮件的地址 */
            Uri uri = Uri.parse("mailto:3gdci@mail.com");
            /* 创建 Intent */
            Intent it = new Intent(Intent.ACTION_SENDTO, uri);
            /* 设置邮件的主题 */
            it.putExtra(android.content.Intent.EXTRA_SUBJECT, "数据备份");
            /* 设置邮件的内容 */
            it .putExtra(android.content.Intent.EXTRA_TEXT, "本次计数:
"+miCount);

            /* 开启 */
            startActivity(it);

            this.finish();
            return true;
        }
        return super.onKeyDown(keyCode, event);
    }
}
```

代码运行效果如图 6-6 所示。

图　6-6

既然可以在网络上保存文件，就可以从网络上读取文件，有兴趣的读者请自行研究，这里不再赘述。

6.5 Android 数据库编程

前面介绍了在 Android 平台中存储数据的 3 种方式，可以看出，这 3 种方式都是用于存储简单的、数据量较小的数据，如果要存储、管理以及升级维护（比如实现一个理财工具）大量数据，可能需要随时添加、查看或更新数据，上述 3 种方式将不再适应。Android 系统考虑到这个问题，提供了 SQLite 数据库，专门处理数据量较大的数据。它在数据的存储、管理、维护等各个方面都更加合理，功能更加强大。很多系统自带的应用采用 SQLite 数据库，比如电话本。本节将详细地介绍 SQLite 数据库在 Adroid 系统中的应用。

6.5.1 SQLite 简介

SQLite 诞生于 2000 年 5 月，它是一款轻型数据库，是嵌入式的，目前在很多嵌入式产品中使用。它占用资源非常少，在嵌入式设备中只需要几百 KB 内存就足够。也许这正是 Android 系统采用 SQLite 数据库的原因之一。

SQLite 数据库是 D.Richard Hipp 用 C 语言编写的开源嵌入式数据库，支持的数据库大小为 2TB，具有下述特征。

1．轻量级

SQLite 与 C/S 模式的数据库软件不同，它是进程内的数据库引擎，因此不存在数据库的客户端和服务器。使用 SQLite，一般只需要带上它的一个动态库，就可以享受其全部功能，而且动态库的尺寸也非常小。

2．独立性

SQLite 数据库的核心引擎本身不依赖第三方软件，使用它不需要“安装”，所以部署时能够省去不少麻烦。

3．隔离性

SQLite 数据库中所有的信息（比如表、视图、触发器等）都包含在一个文件内，方便管理和维护。

4．跨平台

SQLite 数据库支持大部分操作系统，除了在电脑上使用的操作系统之外，还可以运行很多手机操作系统，比如 Android、Windows Mobile、Symbin、Plam 等。

5．多语言接口

SQLite 数据库支持很多语言编程接口，比如 C/C++、Java、Python、dotNet、Ruby、Perl 等，得到更多开发者的喜爱。

6．安全性

SQLite 数据库通过其独占性和共享锁来实现独立事务处理。这意味着多个进程可以在同一时间、同一数据库读取数据，但只有一个可以写入数据。在某个进程或线程向数据库执行写操作之前，必须获得独占锁定。在发出独占锁后，其他的读或写操作将不再发生。

SQLite 数据库的优点很多，由于篇幅所限，仅列举这些。如果要了解更多信息，请浏览 SQLite 官方网站（http://www.sqlite.org/）。下面介绍在 Android 系统中如何使用 SQLite 数据库。

6.5.2 SQLite 编程详解

SQLite 数据库功能非常强大，使用起来非常方便。SQLite 数据库的一般操作包括：创建数据库、打开数据库、创建表、向表中添加数据、从表中删除数据、修改表中的数据、关闭数据库、删除指定表、删除数据库和查询表中的某条数据，分述如下。

1．创建和打开数据库

在 Android 中创建和打开数据库可以使用 openOrCreateDatabase 方法，它自动检测是否存在这个数据库。如果存在，则打开；如果不存在，则创建一个数据库。创建成功，返回一个 SQLiteDatabase 对象，否则抛出异常 FileNotFoundException。下面创建一个名为 dci.db 的数据库，并返回一个 SQLiteDatabase 对象 mySQLiteDatabase，代码如下所示：

```
    /* 数据库对象 */
private SQLiteDatabase mySQLiteDatabase  = null;
   /* 数据库名 */
private final static String  DATABASE_NAME = "dci.db";
   // 打开已经存在的数据库
mySQLiteDatabase=this.openOrCreateDatabase(DATABASE_NAME,MODE_PRIV
ATE,null);
```

2．创建表

一个数据库中可以包含多个表，每一条数据都保存在一个指定的表中。要创建表，通过 execSQL 方法执行一条 SQL 语句。execSQL 能够执行大部分 SQL 语句。下面创建一个名为 student 且包含 3 个字段的表，代码如下所示：

```
/* 表名 */
   private final static StringTABLE_NAME      = "student";
/* 表中的字段 */
   private final static StringSTU_ID    = "_id";
   private final static StringSTU_NUM       = "num";
   private final static StringSTU_DATA  = "data";
/* 创建表的 sql 语句 */
   private final static StringCREATE_TABLE  =  "CREATE  TABLE  "  +
TABLE_NAME + " (" + STU_ID + " INTEGER PRIMARY KEY," + STU_NUM    +
" INTERGER,"+ STU_DATA + " TEXT)";
```

3．向表中添加一条数据

可以使用 insert 方法来添加数据，但是 insert 方法要求把数据打包到 ContentValues 中。ContentValues 其实就是一个 Map，Key 值是字段名称，Value 值是字段的值。通过 ContentValues 中的 put 方法把数据放到 ContentValues 中，然后插入到表中，代码如下所示：

```
ContentValues cv = new ContentValues();
   cv.put(STU_NUM, miCount);
   cv.put(STU_DATA, "测试数据库数据");
   /* 插入数据 */
   mySQLiteDatabase.insert(TABLE_NAME, null, cv);
```

在这里，同样使用 execSQL 方法执行一条“插入”的 SQL 语句，代码如下所示：

```
String INSERT_DATA="INSERT INTO student(_id,num,data) values(1,1,"
通过 SQL 语句插入")";
mySQLiteDatabase.execSQL(INSERT_DATA);
```

4．从表中删除数据

删除数据使用 delete 方法。下面删除字段_id 等于 1 的数据，代码如下所示：

```
mySQLiteDatabase.delete("dci.db","WHERE _id="+1,null);
```

通过 execSQL 方法执行 SQL 语句删除数据，代码如下所示：

```
/* 删除数据 */
    String DELETE_DATA= "DELETE FROM student WHERE _id=1"
mySQLiteDatabase.execSQL(DELETE_DATA);
```

5．修改表中数据

如果添加数据后发现有误，需要修改，可以使用 update 方法。下面修改 num 值为 0 的数据，代码如下所示：

```
ContentValues cv = new ContentValues();
cv.put(STU_NUM, 3);
cv.put(STU_DATA, "修改后的数据");
/* 更新数据 */
mySQLiteDatabase.update(TABLE_NAME, cv, STU_NUM + "=" + Integer.
toString(0), null);
```

6．关闭数据库

关闭数据库很重要，也是用户容易忘记的。关闭的方法很简单，直接使用 SQLiteDatabase 的 close 方法，代码如下所示：

```
mySQLiteDatabase.close();
```

7．删除指定表

使用 execSQL 方法删除指定表，代码如下所示：

```
mySQLiteDatabase.execSQL("DROP TABLE student");
```

8．删除数据库

要删除一个数据库，直接使用 deleteDatabase 方法，代码如下所示：

```
this.deleteDatabase("dci.db");
```

9．查询表中的某条数据

在 Android 中查询某条数据，通过 Cursor 来实现。使用 SQLiteDatabase.query()方法时，得到一个 Cursor 对象，Cursor 指向的就是每一条数据。Cursor 类常见方法如表 6-1 所示。

表 6-1

方　法	说　明
move	以当前位置为参考，将 Cursor 移动到指定的位置。成功，返回 true；失败，返回 false
moveToPosition	将 Cursor 移动一个位置。成功，返回 true；失败，返回 false
moveToNext	将 Cursor 向前移动一个位置。成功，返回 true；失败，返回 false
moveToLast	将 Cursor 向后移动一个位置。成功，返回 true；失败，返回 false
moveToFirst	将 Cursor 移动到第一行。成功，返回 true；失败，返回 false
isBeforeFirst	返回 Cursor 是否指向第一项数据之前
isAfterLast	返回 Cursor 是否指向最后一项数据之后
isClosed	返回 Cursor 是否关闭
isFirst	返回 Cursor 是否指向第一项数据
isLast	返回 Cursor 是否指向最后一项数据
isNull	返回指定位置是否为 NULL
getCount	返回总的数据项数
getInt	返回当前行中指定索引的数据

用 Cursor 来查询数据库中的数据，代码如下所示：

```
Cursor cur= mySQLiteDatabase.rawQuery("SELECT * FROM student",null);
if(cur!=null){
   if(cur.moveToFirst()){
     do{
        int numColumn =cur.getColumIndex("num");
        Int num = cur.getInt(numColumn);
      }while(cur.moveToNext());
   }
}
```

上面介绍了 SQLite 数据库的基本操作，现在来实现一个完整的数据库操作。首先运行项目，初始化数据库（创建数据库、创建表）。然后，单击左方向键，向表中插入一条数据；按右方向键，删除一条数据，如图 6-7 所示；按数字键 1，修改表中指定的一条数据。

如图 6-8 所示，按数字键 2，删除一个表；按数字键 3，删除数据库。

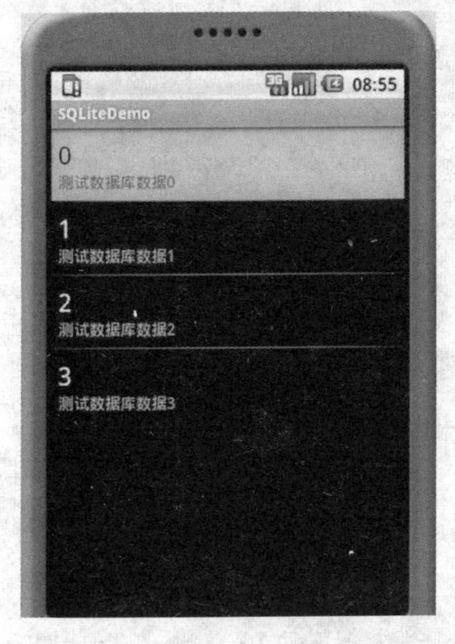

图　6-7

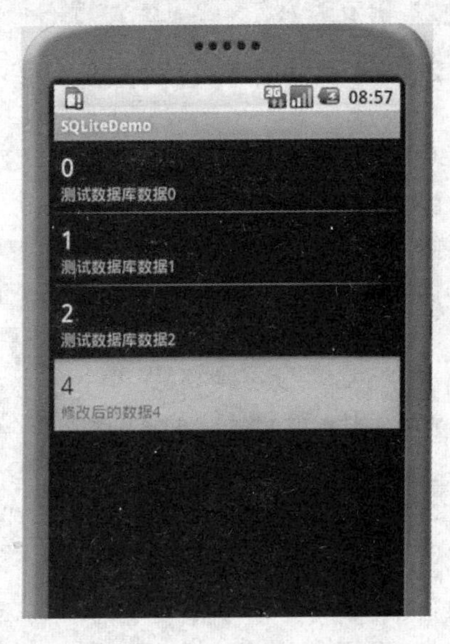

图　6-8

在图 6-7 中，通过 ListView 显示数据库中的数据。该例创建了一个具有 3 个字段的表，分别是_id、num 和 data。其中，_id 通过系统自动赋值（INTEGER PRIMARY KEY），其他两个字段的值通过 ContentValues 使用 insert 插入，代码如下所示：

```
package com.redarmy.sd;

import android.app.Activity;
import android.content.ContentValues;
import android.database.Cursor;
import android.database.sqlite.SQLiteDatabase;
import android.graphics.Color;
import android.os.Bundle;
import android.view.KeyEvent;
import android.widget.LinearLayout;
```

```
import android.widget.ListAdapter;
import android.widget.ListView;
import android.widget.SimpleCursorAdapter;

public class ActivityMain extends Activity
{
    private static int          miCount          = 0;

    /* 数据库对象 */
    private SQLiteDatabase      mSQLiteDatabase  = null;

    /* 数据库名 */
    private final static String DATABASE_NAME = "Examples_06_05.db";
    /* 表名 */
    private final static String TABLE_NAME    = "table1";

    /* 表中的字段 */
    private final static String TABLE_ID      = "_id";
    private final static String TABLE_NUM     = "num";
    private final static String TABLE_DATA    = "data";

    /* 创建表的sql语句 */
    private final static String CREATE_TABLE  =  "CREATE  TABLE  "  +
TABLE_NAME + " (" + TABLE_ID + " INTEGER PRIMARY KEY," + TABLE_NUM +
" INTERGER,"+ TABLE_DATA + " TEXT)";

    /* 线性布局 */
    LinearLayout            m_LinearLayout   = null;
    /* 列表视图-显示数据库中的数据 */
    ListView                m_ListView    = null;
    /** Called when the activity is first created. */
    @Override
    public void onCreate(Bundle savedInstanceState)
    {
        super.onCreate(savedInstanceState);

        /* 创建LinearLayout布局对象 */
        m_LinearLayout = new LinearLayout(this);
        /* 设置布局LinearLayout的属性 */
        m_LinearLayout.setOrientation(LinearLayout.VERTICAL);

    m_LinearLayout.setBackgroundColor(android.graphics.Color.BLACK);
```

```
        /* 创建 ListView 对象 */
        m_ListView = new ListView(this);
        LinearLayout.LayoutParams    param    =    new    LinearLayout.
LayoutParams(LinearLayout.LayoutParams.FILL_PARENT,   LinearLayout.
LayoutParams.WRAP_CONTENT);
        m_ListView.setBackgroundColor(Color.BLACK);

        /* 添加 m_ListView 到 m_LinearLayout 布局 */
        m_LinearLayout.addView(m_ListView, param);

        /* 设置显示 m_LinearLayout 布局 */
        setContentView(m_LinearLayout);

        // 打开已有的数据库
        mSQLiteDatabase  =  this.openOrCreateDatabase(DATABASE_NAME,
MODE_PRIVATE, null);

        // 获取数据库 Phones 的 Cursor
        try
        {
            /* 在数据库 mSQLiteDatabase 中创建一个表 */
            mSQLiteDatabase.execSQL(CREATE_TABLE);
        }
        catch (Exception e)
        {
            UpdataAdapter();
        }
    }
    public boolean onKeyUp(int keyCode, KeyEvent event)
    {
        switch (keyCode)
        {
            case KeyEvent.KEYCODE_DPAD_LEFT:
                AddData();
                break;
            case KeyEvent.KEYCODE_DPAD_RIGHT:
                DeleteData();
                break;
            case KeyEvent.KEYCODE_1:
                UpData();
                break;
```

```
            case KeyEvent.KEYCODE_2:
                DeleteTable();
                break;
            case KeyEvent.KEYCODE_3:
                DeleteDataBase();
                break;
        }
        return true;
    }
    /* 删除数据库 */
    public void DeleteDataBase()
    {
        this.deleteDatabase(DATABASE_NAME);
        this.finish();
    }
    /* 删除一个表 */
    public void DeleteTable()
    {
        mSQLiteDatabase.execSQL("DROP TABLE " + TABLE_NAME);
        this.finish();
    }
    /* 更新一条数据 */
    public void UpData()
    {
        ContentValues cv = new ContentValues();
        cv.put(TABLE_NUM, miCount);
        cv.put(TABLE_DATA, "修改后的数据" + miCount);
        /* 更新数据 */
        mSQLiteDatabase.update(TABLE_NAME, cv, TABLE_NUM + "=" +
Integer.toString(miCount - 1, null);
        UpdataAdapter();
    }
    /* 向表中添加一条数据 */
    public void AddData()
    {
        ContentValues cv = new ContentValues();
        cv.put(TABLE_NUM, miCount);
        cv.put(TABLE_DATA, "测试数据库数据" + miCount);
        /* 插入数据 */
        mSQLiteDatabase.insert(TABLE_NAME, null, cv);
        miCount++;
        UpdataAdapter();
```

```
    }
    /* 从表中删除指定的一条数据 */
    public void DeleteData()
    {
        /* 删除数据 */
        mSQLiteDatabase.execSQL("DELETE FROM " + TABLE_NAME + " WHERE
_id=" + Integer.toString(miCount));
        miCount--;
        if (miCount < 0)
        {
            miCount = 0;
        }
        UpdataAdapter();
    }
    /* 更新视图显示 */
    public void UpdataAdapter()
    {
        // 获取数据库 Phones 的 Cursor
        Cursor cur = mSQLiteDatabase.query(TABLE_NAME, new String[]
{ TABLE_ID, TABLE_NUM, TABLE_DATA }, null, null, null, null, null);
        miCount = cur.getCount();
        if (cur != null && cur.getCount() >= 0)
        {
            // ListAdapter 是 ListView 和后台数据的桥梁
            ListAdapter adapter = new SimpleCursorAdapter(this,
            // 定义 List 中每一行的显示模板
                // 表示每一行包含两个数据项
                android.R.layout.simple_list_item_2,
                // 数据库的 Cursor 对象
                cur,
                // 从数据库的 TABLE_NUM 和 TABLE_DATA 两列中取数据
                new String[] { TABLE_NUM, TABLE_DATA },
                // 与 NAME 和 NUMBER 对应的 Views
                new int[] { android.R.id.text1, android.R.id.text2 });
            /* 将 adapter 添加到 m_ListView 中 */
            m_ListView.setAdapter(adapter);
        }
    }
    /* 按键事件处理 */
    public boolean onKeyDown(int keyCode, KeyEvent event)
    {
        if (keyCode == KeyEvent.KEYCODE_BACK)
```

```
        {
            /* 退出时，不要忘记关闭 */
            mSQLiteDatabase.close();
            this.finish();
            return true;
        }
        return super.onKeyDown(keyCode, event);
    }
}
```

6.6 Content Provider

Content Providers 用于存取数据，可以使所有应用程序访问这些数据。Content Provider 是应用程序之间共享数据的唯一方式。不存在所有 Android 包都能够访问的共用存储区。

Android 系统中有一些常用数据类型的 Content Provider（音频、视频、图像、个人通讯录等），在 android.provider 包中可以找到。可以查询这些 Provider 包含的数据（对某些数据来说，必须获得必要的授权 root 权限）。

本节介绍怎样使用 Content Provider。介绍其基本概念之后，将探讨如何查询 Content Provider、修改 Provider 控制的数据，以及创建 Content Provider。

6.6.1 数据模型

Content Provider 把数据作为数据库模型上的单一数据表呈现出来。在数据表中，每一行是一条记录，每一列代表某一特定类型的值。例如，联系人信息及其电话号码输出如下：

_ID	Number	Number_key	Lable	Name	Type
13	(425) 555 6677	425 555 6677	Kirkland office	Bully Pulpit	TYPE_WORK
44	(212) 555-1234	212 555 1234	NY apartment	Alan Vain	TYPE_HOME
45	(212) 555-6657	212 555 6657	Downtown office	Alan Vain	TYPE_MOBILE
53	201.555.4433	201 555 4433	Love Nest	Rex Cars	TYPE_HOME

每条记录包含了_ID 域，它在表中唯一标识该记录。ID 也用于在相关表中匹配激励。例如，在一个表中查找一个人的电话号码，在另一个表中获取其照片。

查询返回一个 Cursor 对象。Cursor 可以从一条记录移到另一条记录，一列一列地读取每个字段的数据。读取每种类型的数据有特定的方法，所以，如果要读取一个字段，必须知道该字段包含的数据的类型。

6.6.2 URI

每个 Content Provider 都会对外提供一个公共的 URI（包装成 URI 对象）。URI 唯一确定该 Content Provider 的数据集。控制多个数据集（多个表）的 Content Provider 为每一个数据集提供独立的 URI。Provider 的 URI 都以字符串 content://开头。content:表示这些数据被 Content Provider 控制。

如果自定义一个 Content Provider，应同时为其 URI 定义一个常量，这不仅可以简化客户端代码，而且使将来的升级更便捷。对于所有与 Android 平台一同发布的 Provider，系统都为其定义了 CONTENT_URI 常量。例如，匹配电话号码和联系人表的 URI，以及保存联系人照片表的 URI，如下所示：（这两个表都由 Contacts 的 Content Provider 控制）：

```
android.provider.Contacts.Phones.CONTENT_URI
   android.provider.Contacts.Photos.CONTENT_URI
```

类似地，通话记录表的 URI 和日程表的 URI 分别是：

```
android.provider.CallLog.Calls.CONTENT_URI
   android.provider.Calendar.CONTENT_URI
```

在所有与 Content Provider 互动的过程中，都可以使用这个 URI 常量。每个 ContentResolver 都把 URI 作为其第一个参数，它决定了 ContentResolver 将与哪一个 Provider 对话，该 Provider 对应的是哪一个数据表。

6.6.3　查询

查询 Content Provider 需要下述三个信息：

（1）标识 Provider 的 URI。

（2）想要获取的数据字段的名称。

（3）这些字段的数据类型。

如果要访问一条特殊记录，也需要其 ID。

通过调用 ContentResolver.query()函数或者 Activity.managedQuery()函数来查询 Content Provider。这两个函数携带相同的参数，并且均返回一个 Cursor 对象。但是，managedQuery()函数使 Activity 管理 Cursor 的生命周期。被管理的 Cursor 负责处理所有具体事务。例如，当 Activity 暂停的时候，Cursor 负责把自己卸载；当 Activity 重新启动的时候，重新查询。通过调用 Activity.startManagingCursor()函数，可以让 Activity 管理一个没有被管理的 Cursor 对象。

传给 query() 或 managedQuery()的第一个参数是 Provider 的 URI。CONTENT_URI 常量标识了一个特定的 ContentProvider 及其数据集。

在 URI 后面附上记录的_ID 值，就可以把查询限制到一条记录上，即把匹配 ID 值的一个字符串作为 URI 路径部分的最后一段。例如，若 ID 是 23，那么 URI 的形式为：

```
content://. . . ./23
```

通过 ContentUris.withAppendedId()和 Uri.withAppendedPath()这样的辅助方法，很容易把一个 ID 附加到 URI 上。这两个函数都是静态方法，返回一个附加了 ID 的 URI 对象。例如，若想在联系人数据库中查找第 3 条记录，构造一个如下所示的查询：

```
import android.provider.Contacts.People;
import android.content.ContentUris;
import android.net.Uri;
import android.database.Cursor;
Uri myPerson = ContentUris.withAppendedId(People.CONTENT_URI, 3);
Uri myPerson = Uri.withAppendedPath(People.CONTENT_URI, "3");
Cursor cur = managedQuery(myPerson, null, null, null, null);
```

query()和 managedQuery()函数的其他参数更详细地界定了查询，如下所述。

（1）返回的列的名称。null 表示返回所有列；否则，只返回列出名字的列。所有 Android 平台提供的 Content Providers 都为自己的列定义了常量。例如，android.provider.Contacts.Phones 类为 Phone 数据表中的列名定义了常量_ID、NUMBER、NUMBER_KEY、NAME 等。

（2）过滤器的格式如同 SQL 的 Where 语句，它定义了返回哪些行。如果 URI 没有显示产生某一条记录，null 值表示返回查到的所有行。

（3）选择参数。

（4）返回行的排序，格式定义如同 SQL 的 ORDER BY 语句。null 值表示返回的记录按照其在数据表中的缺省值排序，可能是无序的。

下面是一个获取联系人姓名及其主要电话号码等信息的例子。

```
import android.provider.Contacts.People;
import android.database.Cursor;
String[] projection = new String[] {
                        People._ID,
                        People._COUNT,
                        People.NAME,
                        People.NUMBER
                      };
Uri contacts =  People.CONTENT_URI;
Cursor managedCursor = managedQuery(contacts,
                     projection,
                     null,
                     null,
                     People.NAME + "a");
```

这次查询从通信录 Content Provider 的人员表中获取数据，得到每位联系人的姓名、电话号码和唯一的记录 ID，它报告了记录的数量，并将其作为每条记录的__COUNT 字段返回。

列名的常量在不同的接口中定义。在 BaseColumns 中定义了_ID 和_COUNT，在 PeopleColumns 中定义了 NAME，在 PhoneColumns 中定义了 NUMBER。Contacts.People 类实现了所有接口，因此上述代码能够通过类名引用它们。

6.6.4 修改记录

可以使用 ContentResolver.update()方法来修改数据，代码如下所示：

```
private void updateRecord(int recNo, String name) {
   Uri uri = ContentUris.withAppendedId(People.CONTENT_URI, recNo);
   ContentValues values = new ContentValues();
   values.put(People.NAME, name);
   getContentResolver().update(uri, values, null, null);
}
```

调用上述方法更新指定记录的代码如下所示：

```
updateRecord(10, "XYZ");   //更改第 10 条记录的 name 字段值为"XYZ"
```

6.6.5 添加记录

可以调用 ContentResolver.insert()方法添加记录。该方法接收一个要添加的记录的目标 URI，以及一个包含了新记录值的 Map 对象。调用后的返回值是新记录的 URI，包含记录号。

上述例子都是基于联系人信息簿这个标准的 Content Provider，下面创建一个 insertRecord()方法为联系人信息簿添加数据，代码如下所示：

```
**
* 新增联系人
*
* @param name
* @param phoneNo
*/
/*
* private void insertRecords(String name, String phoneNo) { Content
Resolver
* resolver = this.getContentResolver(); ContentValues values = new
* ContentValues(); values.put(People.NAME, name); Uri uri =
* getContentResolver().insert(People.CONTENT_URI, values); Log.d("ANDROID",
* uri.toString()); Uri numberUri = Uri.withAppendedPath(uri,
* People.Phones.CONTENT_DIRECTORY); values.clear();
* values.put(Contacts.Phones.TYPE, People.Phones.TYPE_MOBILE);
* values.put(People.NUMBER, phoneNo); resolver.insert(numberUri, values); }
*/
```

上述代码调用 insertRecords(name, phoneNo)方法向联系人信息簿添加联系人姓名和电话号码。

6.6.6　删除记录

Content Provider 中的 getContextResolver.delete()方法用来删除记录。下述代码用于删除设备上所有的联系人信息：

```
 private void deleteRecords() {
   Uri uri = People.CONTENT_URI;
   getContentResolver().delete(uri, null, null);
}
```

也可以指定 Where 条件语句来删除特定的记录：

```
getContentResolver().delete(uri,"NAME="+ "'XYZ XYZ'", null);
```

执行上述代码，将删除 name 为'XYZ XYZ'的记录。

6.6.7　创建 Content Provider

创建 Content Provider 的操作步骤如下所述：

（1）创建一个继承了 ContentProvider 父类的类。

（2）定义一个名为 CONTENT_URI，并且是 public static final 的 URI 类型的类变量，必须为其指定唯一的字符串值，最好是类的全名，例如，

```
public static final Uri CONTENT_URI = Uri.parse("content://com.google.
android.MyContentProvider");
```

（3）创建数据存储系统。大多数 Content Provider 使用 Android 文件系统或 SQLite 数据库来存储数据，也可以以任何想要的方式来存储。

（4）定义要返回给客户端的数据列名。如果正在使用 Android 数据库，则数据列的使用方式和其他数据库一样。但是，必须为其定义一个名为_id 的列，表示每条记录的唯一性。

（5）如果要存储字节型数据，比如位图文件等，那么保存该数据的数据列其实是一个保存文件的 URI 字符串，客户端通过它来读取对应的文件数据。处理这种数据类型的 Content Provider 需要实现一个名为_data 的字段，其中列出了该文件在 Android 文件系统的精确路径。该字段不仅供客户端使用，也可以供 ContentResolver 使用。客户端可以调用 ContentResolver.openOutputStream() 方法来处理该 URI 指向的文件资源。如果是 ContentResolver 本身，由于其权限比客户端高，所以它能直接访问该数据文件。

（6）声明 public static String 型的变量，用于指定从游标处返回的数据列。

（7）查询返回一个 Cursor 类型的对象。所有执行写操作的方法，如 insert()、update() 及 delete()都将被监听。可以通过使用 ContentResover().notifyChange()方法来通知监听器关于数据更新的信息。

（8）在 AndroidMenifest.xml 中使用<provider>标签设置 Content Provider。

（9）如果要处理的是一种比较新的数据类型，必须先定义一个新的 MIME 类型，供 ContentProvider.geType(url)返回。MIME 类型有两种形式：一种是指定的单个记录，另一种是多条记录。常用格式如下所示：

① vnd.android.cursor.item/vnd.yourcompanyname.contenttype（单个记录的 MIME 类型）：

比如一个请求列车信息的 URI，如 content://com.example.transportationprovider/trains/122 可能返回 MIME 类型 typevnd.android.cursor.item/vnd.example.rail。

② vnd.android.cursor.dir/vnd.yourcompanyname.contenttype （多个记录的 MIME 类型）

比如一个请求所有列车信息的 URI，如 content://com.example.transportationprovider/trains 可能返回 MIME 类型 vnd.android.cursor.dir/vnd.example.rail。

下列代码将创建一个 Content Provider，仅存储，并显示所有的用户名称（使用 SQLLite 数据库存储数据）：

```
package com.wissen.testApp;

public class MyUsers {
   public static final String AUTHORITY = "com.wissen. MyContentProvider";

   // BaseColumn 类中已经包含了 _id 字段
  public static final class User implements BaseColumns {
       public static final Uri CONTENT_URI  = Uri.parse("content://
com.wissen.MyContentProvider");
       // 表数据列
    public static final String  USER_NAME  = "USER_NAME";
   }
}
```

上述类中定义了 Content Provider 的 CONTENT_URI 以及数据列。下面将基于这些类定义实际的 Content Provider 类。

```
package com.wissen.testApp.android;

public class MyContentProvider extends ContentProvider {
   private SQLiteDatabase    sqlDB;
```

```
    private DatabaseHelper   dbHelper;
    private static final String  DATABASE_NAME    = "Users.db";
    private static final int       DATABASE_VERSION       = 1;
    private static final String TABLE_NAME   = "User";
    private static final String TAG = "MyContentProvider";

    private static class DatabaseHelper extends SQLiteOpenHelper {
        DatabaseHelper(Context context) {
            super(context, DATABASE_NAME, null, DATABASE_VERSION);
        }

        @Override
        public void onCreate(SQLiteDatabase db) {
            //创建用于存储数据的表
        db.execSQL("Create table " + TABLE_NAME + "( _id INTEGER PRIMARY
KEY AUTOINCREMENT, USER_NAME TEXT);");
        }

        @Override
        public void onUpgrade(SQLiteDatabase db, int oldVersion, int
newVersion) {
            db.execSQL("DROP TABLE IF EXISTS " + TABLE_NAME);
            onCreate(db);
        }
    }

    @Override
    public int delete(Uri uri, String s, String[] as) {
        return 0;
    }

    @Override
    public String getType(Uri uri) {
        return null;
    }

    @Override
    public Uri insert(Uri uri, ContentValues contentvalues) {
        sqlDB = dbHelper.getWritableDatabase();
        long rowId = sqlDB.insert(TABLE_NAME, "", contentvalues);
        if (rowId > 0) {
            Uri rowUri = ContentUris.appendId(MyUsers.User.CONTENT_URI.
```

```
buildUpon(), rowId). build();
            getContext().getContentResolver().notifyChange(rowUri,null);
            return rowUri;
        }
        throw new SQLException("Failed to insert row into " + uri);
    }

    @Override
    public boolean onCreate() {
        dbHelper = new DatabaseHelper(getContext());
        return (dbHelper == null) ? false : true;
    }

    @Override
    public Cursor query(Uri uri, String[] projection, String selection,
String[] selectionArgs, String sortOrder) {
        SQLiteQueryBuilder qb = new SQLiteQueryBuilder();
        SQLiteDatabase db = dbHelper.getReadableDatabase();
        qb.setTables(TABLE_NAME);
        Cursor c = qb.query(db, projection, selection, null, null, null,
sortOrder);
        c.setNotificationUri(getContext().getContentResolver(), uri);
        return c;
    }

    @Override
    public int update(Uri uri, ContentValues contentvalues, String s,
String[] as) {
        return 0;
    }
}
```

一个名为 MyContentProvider 的 Content Provider 创建完成了，它用于从 SQLite 数据库中添加和读取记录。

Content Provider 的入口需要在 AndroidManifest.xml 中配置，代码如下所示：

```
<provider android:name="MyContentProvider" android:authorities="com.
wissen.MyContentProvider" />
```

然后，就可以使用这个定义好的 Content Provider，代码如下所示：

```
package com.wissen.testApp;

public class MyContentDemo extends Activity {
    @Override
```

```
    protected void onCreate(Bundle savedInstanceState) {
        super.onCreate(savedInstanceState);
        insertRecord("MyUser");
        displayRecords();
    }

    private void insertRecord(String userName) {
        ContentValues values = new ContentValues();
        values.put(MyUsers.User.USER_NAME, userName);
        getContentResolver().insert(MyUsers.User.CONTENT_URI,values);
    }

    private void displayRecords() {
        String columns[] = new String[] { MyUsers.User._ID, 
MyUsers. User.USER_NAME };
        Uri myUri = MyUsers.User.CONTENT_URI;
        Cursor cur = managedQuery(myUri, columns,null, null, null );
        if (cur.moveToFirst()) {
            String id = null;
            String userName = null;
            do {
                id = cur.getString(cur.getColumnIndex(MyUsers.User._ID));
                userName  =  cur.getString(cur.getColumnIndex(MyUsers.
User.USER_NAME));
                Toast.makeText(this, id + " " + userName, 
Toast.LENGTH_ LONG).show();
            } while (cur.moveToNext());
        }
    }
}
```

上面的类将先向数据库添加一条用户数据，然后显示数据库中所有的用户数据。

第 7 章　Service 应用

7.1　Service 概述

与 Activity 相反，Service 没有可见的用户界面，但是能长时间地在后台运行。因此可以这样理解，Service 是具有一段较长生命周期且没有用户界面的程序。

为什么需要长时间运行在后台的 Service?以音乐播放器为例，用户可能在播放音乐的同时要编辑短信或者浏览网页，因此音乐播放器不可能长时间处于前台。为了让音乐一直播放下去，需要将播放音乐的任务放到后台。这样，即使音乐播放器不再显示，用户依然可以听到音乐，所以需要这样的机制——长时间在后台运行的 Service。

Service 与 Activity 及其他组件（BroadcastReceiver 和 ContentProvider）一样，运行在应用程序进程的主线程中，因此 Service 不会阻塞其他组件和用户界面。

7.2　Service 的生命周期

每个 Service 需要像 Activity 一样，在所属包内的 AndroidMainfest.xml 由对应的<Service>标签来声明。若需要指定使用当前 Service 的权限，要在<service>标签内部加上权限标签<uses-permission>。创建 Service 需要创建继承自 android.app.Service 的类，然后重新实现在 Service 各个状态要回调的方法。相比 Activity，Service 的状态回调方法只有三个，分别是 onCreate()、onStart()和 onDestroy()。Service 不能自启动，必须通过 Context 对象（如一个 Activity）调用 startService()或 bindService()方法来启动。采用这两种方法启动的 Service 的生命周期是不同的，如图 7-1 所示。

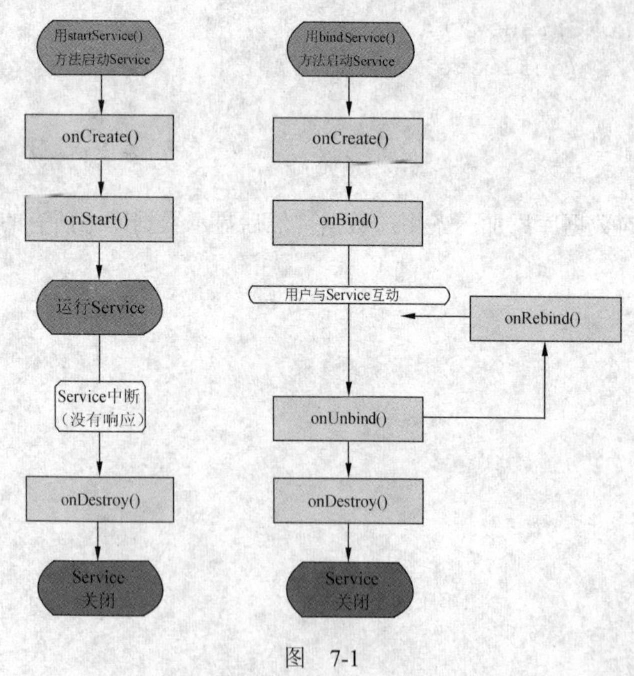

图　7-1

1．调用 startService()方法

若 Service 没有启动，首先调用该 Service 的 onCreate()方法，然后调用 onStart()方法。若 Service 已经启动，直接调用 onStart()方法。通过 startService()启动后，Service 通过 Context 对象调用 stopService()来关闭；也可以通过 Service 自身调用 stopSelf()或 stopSelfResult()来关闭。关闭之前调用 onDestory()方法。

2．通过 bindService()

使当前的 Context 对象通过 ServiceConnection 对象绑定到指定的 Service。若没有启动 Service，首先调用 Service 的 onCreate()方法，来初始化启动的 Service，然后调用 Service 的 onBind()方法初始化绑定。如果绑定 Service 的 Context 对象被销毁，被绑定的 Service 也会调用 onUnbind()和 onDestory()方法并停止运行。需要注意的是，BroadcastReceiver 是不能绑定服务的。一个绑定服务的 Context 对象还可以通过 unbindService()来取消对服务的绑定。取消时，Service 调用 unbind()方法。若 Service 是通过 bindService()启动的，还将调用 onDestory()方法来停止运行。

下面通过在各个状态的回调方法中加入 Log 信息，介绍 Service 的生命周期和使用方法。

第一步：新建 Android 项目，名为 LifeCycleService。

第二步：修改 main.xml 文件，添加 4 个 Button 按钮，代码如下所示：

```
<?xml version="1.0" encoding="utf-8"?>
<LinearLayout
xmlns:android="http://schemas.android.com/apk/res/android"
   android:orientation="vertical"
   android:layout_width="fill_parent"
   android:layout_height="fill_parent"
   >

   <Button android:id="@+id/start_service"android:layout_width="fill_parent"
       android:layout_height="wrap_content" android:text="Start Service"
android:layout_marginLeft="20dip" android:layout_marginRight="20dip"/>

   <Button android:id="@+id/stop_service" android:layout_width= "fill_parent"
       android:layout_height="wrap_content" android:text="Stop Service"
android:layout_marginLeft="20dip" android:layout_marginRight="20dip"/>

   <Button android:id="@+id/bind_service" android:layout_width= "fill_parent"
       android:layout_height="wrap_content" android:text="Bind Service"
android:layout_marginLeft="20dip" android:layout_marginRight="20dip"/>

   <Button android:id="@+id/unbind_service" android:layout_width= "fill_parent"
       android:layout_height="wrap_content" android:text="Unbind Service"
android:layout_marginLeft="20dip" android:layout_marginRight="20dip"/>
```

```
</LinearLayout>
```

第三步：修改 ActivityMain，获得组件，添加监听，并做相应的 Service 绑定，代码如下所示：

```
package com.redarmy.lifecycle.service;

import android.app.Activity;
import android.content.ComponentName;
import android.content.Intent;
import android.content.ServiceConnection;
import android.os.Bundle;
import android.os.IBinder;
import android.view.View;
import android.view.View.OnClickListener;
import android.widget.Button;

public class ActivityMain extends Activity implements OnClickListener{
    public Button startButton;
    public Button stopButton;
    public Button bindButton;
    public Button unbindButton;

    //第一个空的 ServiceConnection 的对象,用于绑定 Service
     public ServiceConnection conn = new ServiceConnection(){

        @Override
        public void onServiceConnected(ComponentName name, IBinder service){
            // TODO Auto-generated method stub

        }
        @Override
        public void onServiceDisconnected(ComponentName name) {
            // TODO Auto-generated method stub

        }
    };

    @Override
    public void onCreate(Bundle savedInstanceState) {
        super.onCreate(savedInstanceState);
```

```
        setContentView(R.layout.main);
        //得到组件
        startButton = (Button)findViewById(R.id.start_service);
        stopButton = (Button)findViewById(R.id.stop_service);
        bindButton = (Button)findViewById(R.id.bind_service);
        unbindButton = (Button)findViewById(R.id.unbind_service);

        //添加监听
        startButton.setOnClickListener(this);
        stopButton.setOnClickListener(this);
        bindButton.setOnClickListener(this);
        unbindButton.setOnClickListener(this);

    }

    @Override
    public void onClick(View v) {
        int temp = v.getId();
        Intent intent = new Intent(ActivityMain.this,LifeCycle Service.class);
        switch (temp) {
        case R.id.start_service:
            startService(intent);
            break;
        case R.id.stop_service:
            stopService(intent);
            break;
        case R.id.bind_service:
            bindService(intent, conn, BIND_AUTO_CREATE);
            break;
        case R.id.unbind_service:
            unbindService(conn);
            break;
        }

    }
}
```

第四步：新建 LifeCycleService 并继承 Service 重写方法，代码如下所示：

```
package com.redarmy.lifecycle.service;

import android.app.Service;
```

```
import android.content.Intent;
import android.os.IBinder;
import android.util.Log;

public class LifeCycleService extends Service {

    public static final String TAG="Service";
    @Override
    public IBinder onBind(Intent arg0) {
        Log.v(TAG, "Service..............onBind");
        return null;
    }

    @Override
    public void onCreate() {
        super.onCreate();
        Log.v(TAG, "Service..............onCreate");
    }

    @Override
    public void onDestroy() {
        super.onDestroy();
        Log.v(TAG, "Service..............onDestroy");
    }

    @Override
    public void onStart(Intent intent, int startId) {
        super.onStart(intent, startId);
        Log.v(TAG, "Service..............onStart");
    }

    @Override
    public boolean onUnbind(Intent intent) {
        Log.v(TAG, "Service..............onUnbind");
        return super.onUnbind(intent);
    }
}
```

通过上述操作，可以通过 Log 信息观察 Service 的生命周期。

7.3 Service 的使用

本节通过案例讲解如何使用 Service，主要介绍 Local Service，也就是用户的 Service，也可以操作其他应用程序的 Service(如果允许)，这涉及一个比较复杂的技术 InterProcess Communication(IPC)，它提供了在不同进程之间通信的机制。通常情况下，Local 就够用了。

Android 系统中有很多 Service，例如，常见的 AlarmService、GPS Service 等。除此之外，用户可以定制需要的 Service，如音乐播放 Service 等。下面通过两个例子分别介绍 Service 的使用方法。

1．系统自带 Service 案例：定时提醒

第一步：创建一个 Android 项目，名为 remindtime，并且修改布局模板 main.xml 的代码，如下所示：

```
<?xml version="1.0" encoding="utf-8"?>
<LinearLayout
xmlns:android="http://schemas.android.com/apk/res/android"
   android:orientation="vertical"
   android:layout_width="fill_parent"
   android:layout_height="fill_parent"
   >
<TextView
   android:layout_width="fill_parent"
   android:layout_height="wrap_content"
   android:text="@string/hello"
   />

   <Button
   android:id="@+id/remind_time_service"
   android:layout_width="fill_parent"
   android:layout_height="wrap_content"
   android:layout_marginLeft="20dip"
   android:layout_marginRight="20dip"
   android:text="@string/remind_time"/>
  <Button
   android:id="@+id/exit"
   android:layout_width="fill_parent"
   android:layout_height="wrap_content"
   android:text="Exit"
   android:layout_marginLeft="20dip"
   android:layout_marginRight="20dip"/>

</LinearLayout>
```

修改 strings.xml 文件的代码如下所示：

```
<?xml version="1.0" encoding="utf-8"?>
<resources>
   <string name="remind_time">定时提醒</string>
   <string name="hello">精彩实例.定时提醒!</string>
```

```
    <string name="app_name">remindtime</string>
</resources>
```

第二步：单击第一个按钮后，启动一个 Service，等待 15s 后自动停止，或者单击 Exit 按钮停止。代码如下所示：

```
package com.redarmy.time;

import java.util.Calendar;

import android.app.Activity;
import android.app.AlarmManager;
import android.app.PendingIntent;
import android.content.Intent;
import android.os.Bundle;
import android.view.View;
import android.view.View.OnClickListener;
import android.widget.Button;

//http://moazzam-khan.com/blog/?p=157
public class ActivityMain extends Activity implements OnClickListener {
    /** Called when the activity is first created. */
    private static ActivityMain appRef = null;

    private Button b_remind_service, b_exit_service;
    boolean k = false;

    @Override
    public void onCreate(Bundle icicle) {
        super.onCreate(icicle);
        appRef = this;
        setContentView(R.layout.main);
        b_remind_service = (Button) findViewById(R.id.remind_time_ service);
        b_remind_service.setOnClickListener(this);

        b_exit_service = (Button) findViewById(R.id.exit);
        b_exit_service.setOnClickListener(this);
    }

    public static ActivityMain getApp() {
        return appRef;
    }
```

```
    public void btEvent(String data) {
        setTitle(data);
    }

    public void onClick(View arg0) {

        //如果是“开始”按钮被单击
        if (arg0 == b_remind_service) {
            setTitle("Waiting... Alarm=5"); //设置标题
            //跳转 Intent
            Intent intent = new Intent(ActivityMain.this,
RemindReceiver. class);
            //
            PendingIntent p_intent = PendingIntent.getBroadcast(
                    ActivityMain.this, 0, intent, 0);
            //获得 Calendar 对象实例
            Calendar calendar = Calendar.getInstance();
            //设置时间为系统的时间
            calendar.setTimeInMillis(System.currentTimeMillis());

            calendar.add(Calendar.SECOND, 5);
            //得到一个定时的服务器实例
            AlarmManager am = (AlarmManager) getSystemService (ALARM_SERVICE);
            am.set(AlarmManager.RTC_WAKEUP,calendar.getTimeInMillis(),
                    p_intent);
        }

        //如果是“退出”按钮被单击
        if (arg0 == b_exit_service) {
            Intent intent = new Intent(ActivityMain.this,
RemindReceiver. class);
            PendingIntent p_intent = PendingIntent.getBroadcast(
                    ActivityMain.this, 0, intent, 0);
            AlarmManager am = (AlarmManager) getSystemService(ALARM_ SERVICE);
            am.cancel(p_intent);
            finish();
        }
    }
}
```

第三步：上述代码用到了 RemindReceiver.java 的类代码，如下所示：

```
package com.redarmy.time;

import android.content.BroadcastReceiver;
import android.content.Context;
import android.content.Intent;

public class RemindReceiver extends BroadcastReceiver {

    @Override
    public void onReceive(Context context, Intent arg1) {
        context.startService(new Intent(context, NotifyService.class));
    }
}
```

第四步：当 RemindReceiver 收到 Intent 时，用 context.startService 启动服务 NotifyService，代码如下所示：

```
package com.redarmy.time;

import android.app.Service;
import android.content.Intent;
import android.os.IBinder;
import android.widget.Toast;

public class NotifyService extends Service {

    @Override
    public IBinder onBind(Intent arg0) {
        return null;
    }

    @Override
    public void onCreate() {
        super.onCreate();
        ActivityMain atym = ActivityMain.getApp();
        atym.btEvent("from NotifyService");
        Toast.makeText(this, "定时提醒案例成功！",
Toast.LENGTH_ LONG).show();
    }
}
```

第五步：在 AndroidManifest.xml 文件中添加创建的 RemindReceiver 和 NotifyService，代码如下所示：

```
<?xml version="1.0" encoding="utf-8"?>
```

```
<manifest
xmlns:android="http://schemas.android.com/apk/res/android"
    package="com.redarmy.time" android:versionCode="1"
android: versionName="1.0">
    <application android:icon="@drawable/icon" android:label= "@string/
app_name">
        <activity android:name=".ActivityMain" android:label= "@string/
app_name">
            <intent-filter>
                <action android:name="android.intent.action.MAIN" />
                <category android:name="android.intent.category.LAUNCHER"/>
            </intent-filter>
        </activity>
        <receiver android:name="RemindReceiver"></receiver>
        <service android:name="NotifyService"></service>
    </application>
    <uses-sdk android:minSdkVersion="3" />
</manifest>
```

2．自定义 Service 的使用：音乐播放器

按照设想，实现一个音乐播放的案例。单击按钮启动播放 Service 后，在后台播放音乐，执行其他操作，也不会打断播放。

第一步：创建一个 Android 项目，名为 PlayService，并且修改布局模板 main.xml，代码如下所示：

```
<?xml version="1.0" encoding="utf-8"?>
<LinearLayout xmlns:android="http://schemas.android.com/apk/res/android"
    android:orientation="vertical" android:layout_width="fill_parent"
    android:layout_height="fill_parent">
    <TextView android:layout_width="fill_parent"
        android:layout_height="wrap_content"android:text="@string/hello"/>

    <Button android:id="@+id/start" android:layout_width="fill_parent"
        android:layout_height="wrap_content" android:text="Start Play"
        android:layout_marginLeft="20dip" android:layout_marginRight="20dip"/>

    <Button android:id="@+id/stop" android:layout_width="fill_ parent"
        android:layout_height="wrap_content" android:text="Stop Play"
        android:layout_marginLeft="20dip"android:layout_marginRight= "20dip"/>
</LinearLayout>
```

第二步：实现 PlayService.java，代码如下所示：

```
package com.redarmy.service;
```

```
import android.app.Activity;
import android.content.Intent;
import android.os.Bundle;
import android.view.View;
import android.view.View.OnClickListener;
import android.widget.Button;

public class PlayService extends Activity implements OnClickListener {
    /** Called when the activity is first created. */
    @Override
    public void onCreate(Bundle savedInstanceState) {
        super.onCreate(savedInstanceState);
        setContentView(R.layout.main);
        // 获得自己
        Button BtnStart = (Button) findViewById(R.id.start);
        Button BtnStop = (Button) findViewById(R.id.stop);
        // 添加监听
        BtnStart.setOnClickListener(this);
        BtnStop.setOnClickListener(this);

    }

    @Override
    public void onClick(View v) {
        if (v.getId() == R.id.start) {
            startService(new Intent(
                    "com.redarmy.service.PlayService.START_AUDIO_SERVICE"));
        } else if (v.getId() == R.id.stop) {
            stopService(new Intent(
                    "com.redarmy.service.PlayService.START_AUDIO_SERVICE"));
            finish();
        }

    }
}
```

第三步：创建自定义 Service，代码如下所示：

```
package com.redarmy.service;

import android.app.Service;
```

```
import android.content.Intent;
import android.media.MediaPlayer;
import android.os.IBinder;

public class MusicService extends Service {
    private MediaPlayer player;

    @Override
    public IBinder onBind(Intent intent) {
        // TODO Auto-generated method stub
        return null;
    }

    @Override
    public void onStart(Intent intent, int startId) {
        super.onStart(intent, startId);
        player = MediaPlayer.create(this, R.raw.jintian);
        player.start();
    }
    @Override
    public void onDestroy() {
        super.onDestroy();
        player.stop();
    }


}
```

这里声明了一个 MediaPlayer 的对象 Player，然后在其 onStart 中播放指定的 MP3 文件.

第四步：在 AndroidManifest.xml 文件中添加对该 Service 的说明代码，如下所示：

```
<?xml version="1.0" encoding="utf-8"?>
<manifest xmlns:android="http://schemas.android.com/apk/res/android"
    package="com.redarmy.service" android:versionCode="1"
    android:versionName="1.0">
    <application  android:icon="@drawable/icon"  android:label="@string/
app_name">
        <activity android:name=".PlayService" android:label="@string/
app_name">
            <intent-filter>
                <action android:name="android.intent.action.MAIN" />
                <category
```

```
android:name="android.intent.category.LAUNCHER" />
            </intent-filter>
        </activity>
        <service android:name="MusicService">
            <intent-filter>
            <action android:name="com.redarmy.service.PlayService. START_
AUDIO_SERVICE"></action>
                <category android:name="android.intent.category.DEFAULT">
</category>
            </intent-filter>
        </service>
    </application>
    <uses-sdk android:minSdkVersion="3" />
</manifest>
```

可以看到,添加了一个名为 MusicService 的 Service,对应的 Action 为 com.redarmy.service.PlayService.START_AUDIO_SERVICE。运行后就能看到效果。

第 8 章　案例实践:《贪啵虎》游戏设计

8.1　构思

设计灵感来源于《下 100 层》类型的游戏。不同的是在游戏《贪啵虎》中加入了更多的元素，如定时、无敌等道具、怪物；设置了全新的游戏机制，以及具有不同运动类型的挡板；摒弃了以往游戏中的道具模式。为了通过关卡，玩家不但要吃到道具，而且要结合移动的挡板，把握好吃道具的时机。因为某些道具是有时间限制的，且不会再次出现。这为玩家提供了更大的游戏空间，不是一味地乱跳，而是要抓住最好的时机进行操作。因此《贪啵虎》游戏具有更强的可玩性与娱乐性。

8.1.1　游戏的整体框架

本游戏共分 9 个类，如图 8-1 所示.

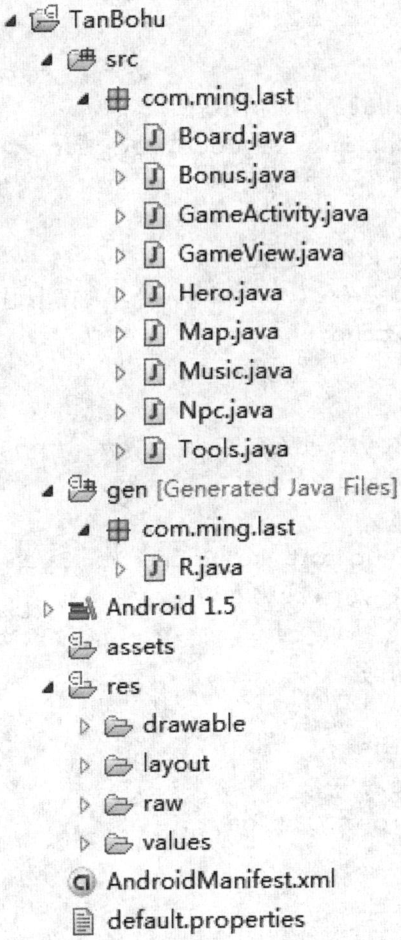

```
TanBohu
  src
    com.ming.last
      Board.java
      Bonus.java
      GameActivity.java
      GameView.java
      Hero.java
      Map.java
      Music.java
      Npc.java
      Tools.java
  gen [Generated Java Files]
    com.ming.last
      R.java
  Android 1.5
  assets
  res
    drawable
    layout
    raw
    values
  AndroidManifest.xml
  default.properties
```

图　8-1

游戏中，所有的绘图与逻辑全部由继承了 SurfaceView 类的 GameView 类完成，它相当于一个总管，管理除 GameActivity 类之外的所有类。GameActivity 类继承自 Activity，负责将 GameView 显示到当前的窗口。Map 类存储游戏中的地图及其他相关数据，供其他类调用。Music 类负责控制游戏中音乐的播放。Npc 类负责生成游戏中的 Npc。Bouns 类负责生成游戏道具。Board 类负责生成游戏中的挡板。Hero 类用于管理主人公。Tools 类封装一些常用的方法，供其他类使用，以避免重复编写代码。

8.1.2 游戏用到的 API

```
java.util.Random;
```

游戏中的背景、主人公与挡板分别有 3 种样式。用 Random 类的 nextInt()方法生成随机数，使得每一关随机地有一幅背景、一个主人公和一种挡板。代码如下所示：

```
public void createRand() {
       tmph = Math.abs(rand.nextInt())%3;
       tmpb = Math.abs(rand.nextInt())%3;
       tmpT = Math.abs(rand.nextInt())%3;
}
android.content.Context;
android.content.SharedPreferences;
android.os.SystemClock;
android.view.KeyEvent;
```

KeyEvent 类，处理按键要用到，如下所示：

```
public boolean onKeyDown(int keyCode, KeyEvent event) {
       if(keyCode == KeyEvent.KEYCODE_BACK ){
           con ++;
           if(con <= 1) {
               Toast.makeText(context, "再次按下退出", 0).show();
               isExit = true;
           }
       }
       return super.onKeyDown(keyCode, event);
   }
   @Override
   public boolean onKeyUp(int keyCode, KeyEvent event) {
       return super.onKeyUp(keyCode, event);
}
android.view.MotionEvent;

android.view.SurfaceHolder;
android.view.SurfaceView;
android.view.SurfaceHolder.Callback;
android.widget.Toast;
android.app.Activity;
android.content.pm.ActivityInfo;
android.os.Bundle;
```

```
android.view.KeyEvent;
android.view.Window;
android.view.WindowManager;
android.content.res.Resources;
android.graphics.Bitmap;
android.graphics.BitmapFactory;
```

BitmapFactory，顾名思义，位图工厂，这个类提供了很多得到 Bitmap 的方法，这里利用其中的一种：BitmapFactory.*decodeResource*(res,id);来返回 Bitmap。

因为在游戏中会加载很多的图片资源,因此在 Tools 类中定义如下方法,用于返回 Bitmap 对象：

```
public Bitmap createBmp(int id){
Bitmap bmp = null;
Resources res = null;
res = ct.getResources();
bmp = BitmapFactory.decodeResource(res, id);
return bmp;
}
android.graphics.Canvas;
android.graphics.Color;
android.graphics.Paint;
android.graphics.Rect;
Rect r = new Rect(int left ,int top,int right ,int bottom );
```

得到矩形对象，在游戏中主要用来实现触屏时检测是否走到屏幕的某一区域。Rect 类中使用方法 contains(x,y)；来检测某点是否在某矩形区域中。

```
public Rect touchRect(int x,int y,int w,int h) {
   return new Rect(x,y,x+w,y+h);
}
在 Tools 类中定义如下方法，用于检测触点是否在某一矩形内
public boolean checkRectTouch(int x,int y,Rect r){
   if(r.contains(x, y)){
      return true;
   }
   return false;
}
android.graphics.Region;
android.media.MediaPlayer;
```

8.2　绘图

8.2.1　游戏 LOGO 的绘制

```
public void drawLogo(Canvas c){
   c.drawColor(Color.WHITE);
```

```
    c.drawBitmap(bmp_logo[logo_index],(scr_w-240)>>1,(scr_h-320)>>1,p);
}
```

首先看方法体的第一句 c.drawColor(Color.WHITE)，其作用是用白色来填充整个屏幕，也就是所谓的清屏。假如不执行这个操作，游戏每循环一次，就会在屏幕上留下上一次绘制的图形，尤其是在图片发生位移时，会特别明显。如果在移动某张图片时发现每次移动都会在图片的后面留下残影，就是由于没有清屏导致的。这是我们不希望见到，所以加上这一句十分必要。

再来看第二句 c.drawBitmap(bmp_logo[logo_index], (scr_w-240)>>1, (scr_h-320)>>1, p);，其作用是将图片 bmp_logo[logo_index]绘制到屏幕的(scr_w-240)>>1,(scr_h-320)>>1 位置。

（1）先来看 bmp_logo[logo_index]。很明显，这是个图片数组。这样写的好处是：假如有多于一张 Logo，只要在逻辑中控制 logo_index 的数值，就可以以规定的时间来播放几张 Logo 图片，很方便。

（2）再来看(scr_w-240)>>1,(scr_h-320)>>1，它是将欲绘制的图片绘制在屏幕的正中（图 8-2）。

图 8-2

<<是左移位运算符，向左移 1 位，相当于乘以 2；>>是右移运算符，向右移 1 位，相当于除以 2。这样书写，使得做乘法或除法运算相当高效。

8.2.2 游戏菜单的绘制

```
public void drawMenu(Canvas c){
    c.drawBitmap(imgmenubk, 0 + 120, 0, p);
    Tools.setClip(c,53 + 120, 260, 42, 17);
    c.drawBitmap(imgMenuZhuanlun, 53 + 120
            , 260 - menuZhuanlunY*17, p);
    Tools.resetClip(c);
    c.drawBitmap(imgMenuBg, 93 + 120, 256, p);
    Tools.setClip(c,103 + 120, 262, 62, 16);
    c.drawBitmap(imgMenuChoose, 103 + 120, 262 - menuPointY, p);
    Tools.resetClip(c);
    c.drawBitmap(imgMenuTiger, 30 + 120, 235, p);
    drawButton(c);
    drawSoft(c,"确定","");
}
```

上述方法与绘制 Logo 的方法基本相同，需要说明的是 Tools.setClip(c,53 + 120, 260, 42, 17);与 Tools.resetClip(c);这两个方法。在 Tools 中定义如下：

```
public static void setClip(Canvas c,int x,int y,int width,int height) {
    c.save();
    c.clipRect(x, y, x+width, y+height,Region.Op.REPLACE);
}
public static void resetClip(Canvas c) {
    c.restore();
}
```

setClip();用来设定剪裁区。c.save();用于记录原来 Canvas 的状态。c.clipRect(x, y, x+width, y+height,Region.Op.REPLACE);用于裁剪图片，需要注意的是后两个参数，如果是从 j2me 转到 Android，开始的时候容易混淆。

注意：前两个参数是左上点坐标，后两个参数是右下点坐标。

resetClip();用来恢复剪裁区。c.restore();用于恢复 Canvas 的状态。需要记住的是，save()与 restore()必须成对出现，也就是说，每次设定剪裁区后，都要恢复。

8.2.3 游戏背景的绘制

```
Tools.setClip(g, 120, 0, 240, 320);
g.drawBitmap(imgGameBg, 0-tmpb*360+Tools.offx, Tools.offy, p);
Tools.resetClip(g);
```

由于背景为整张图片，所以绘制起来比较简单。需要说明的是 Tmpb*360，这里的 Tmpb 是上边产生的[0,2]的随机数，用来随机生成图片的样式。因为每张图片的宽度为 360，所以减去 Tmpb×360，正好是一张图片的宽度，如图 8-3 所示。

图　8-3

Tools.offx 与 Tools.offy 分别为 x 轴与 y 轴的偏移量。因为游戏横向 360 像素，纵向 640 像素，而显示区域为 240×320，涉及摄像机算法，也就是说，屏幕随着 Hero 的移动而相应地滚动。Tools 中代码如下所示：

```
public static void setCamera(Hero hero){
    offy = -(hero.hero_y - ((screenH/5)*3));
    if(offy <= -screenH){
        offy = -screenH;
```

```
    }
    if(offy > 0){
        offy = 0;
    }
    offx = 120 - (hero.hero_x - (screenW>>1));
    if(offx <= 0){
        offx = 0;
    }
    if(offx > 120){
        offx = 120;
    }
}
```

方法中第一句的作用是：当 Hero 的 y 坐标大于屏幕的 3/5 时，开始滚动屏幕中的所有元素。注意，offy 的取值范围是[0,-320]，-320 是用图片的高度减去可视范围的高度得到的，即-320=320-640。对于 offx，同理可得。

8.2.4 游戏元素块的绘制

因为所有的元素块均为对象，所以分别调用其方法进行绘制，代码如下：

```
public static void drawAllBoard(Canvas c,Board[] allb,Paint p){
    for (int i = 0; i < allb.length; i++) {
        if((allb[i].bx + Tools.offx + allb[i].addx) > 120
            && allb[i].bx + Tools.offx < 360
            && (allb[i].by + Tools.offy) < 330
            && (allb[i].by + Tools.offy + 15) > -10){
            if(allb[i].style == 2){
                Tools.setClip(c,allb[i].bx+Tools.offx
                        , allb[i].by + Tools.offy, allb[i].addx, 14);

c.drawBitmap(GameView.imggameBoard,allb[i].bxC+Tools.offx
, allb[i].by + Tools.offy - GameView.tmpT*20, p);
                    Tools.resetClip(c);


            }
            else if(allb[i].style == 3){
Tools.setClip(c,allb[i].bx+Tools.offx,  allb[i].by  +  Tools.offy,
allb[i].addx, 14);
c.drawBitmap(GameView.imggameBoard, allb[i].bxR+Tools.offx, allb[i]. by
+ Tools.offy - GameView.tmpT*20, p);
                Tools.resetClip(c);
            }
            else{
Tools.setClip(c,allb[i].bx+Tools.offx,  allb[i].by  +  Tools.offy,
```

```
allb[i].addx, 14);
c.drawBitmap(GameView.imggameBoard, allb[i].bx+Tools.offx, allb[i]. by
+ Tools.offy - GameView.tmpT*20, p);
                Tools.resetClip(c);
            }
            if(allb[i].style == 4 && allb[i].speed == 3){
Tools.setClip(c,allb[i].bx + 120+Tools.offx, allb[i].by + Tools.offy,
allb[i].addx, 14);
c.drawBitmap(GameView.imggameBoard,  allb[i].bx  +  120+Tools.offx,
allb[i].by + Tools.offy - GameView.tmpT*20, p);
                Tools.resetClip(c);
Tools.setClip(c,allb[i].bx + 240+Tools.offx, allb[i].by + Tools.offy,
allb[i].addx, 14);
c.drawBitmap(GameView.imggameBoard,  allb[i].bx  +  240+Tools.offx,
allb[i].by + Tools.offy - GameView.tmpT*20, p);
                Tools.resetClip(c);
            }
        }
    }
}
//遍历本关所有挡板
for (int i = 0; i < allb.length; i++) {
……
}
//只绘制在可视区域的挡板
if((allb[i].bx + Tools.offx + allb[i].addx) > 120
            && allb[i].bx + Tools.offx < 360
               && (allb[i].by + Tools.offy) < 330
                  && (allb[i].by + Tools.offy + 15) > -10){
……
}
//根据不同的挡板类型，进行不同的绘制
if(allb[i].style == 2){
……
}
//挡板类构造，代码如下
/**
 *
 * @param x    起始横坐标
 * @param y    起始纵坐标
 * @param s    类型 1 表示左边不动，2 表示中间不动，3 表示右边不动
 * @param bc 最大长度/x 方向最大移动到的位置/y 方向最大移动到的位置
```

```
    * @param sp 速度
    * @param add_x 起始长度
    * @param rb 剩余长度/x 方向最小移动到的位置/y 方向最小移动到的位置
    * @param dir 起始方向，1 表示伸长，0 表示收缩/0 表示左或者上/1 表示右或者下
    */
public  Board(int x,int y,int s,int bc,int sp,int add_x,int rb,int dir){
       bx = x;
       by = y;
       oldx = x;
       oldy = y;
       style = s;
       board_max = bc;
       speed = sp;
       board_min = rb;
       isadd = dir;
       addx = add_x;
       bxC = (2*x + add_x - bc)/2;
       bxR = x + add_x - bc;
   }
```

绘制完成的游戏元素块如图 8-4 所示。

图 8-4

8.2.5 游戏人物的绘制

游戏人物如图 8-5 所示，代码如下：

```
public void paint(Canvas c,Paint p) {
   Tools.setClip(c,hero_x+Tools.offx, hero_y + Tools.offy, 32, 32);
   c.drawBitmap(imgHero,  hero_x-(action[state-1][hero_frame])%6  *
   32+  Tools.offx,  hero_y-(action[state-1][hero_frame])/6*32  +
   Tools.offy - GameView.tmph * 64,p);
   Tools.resetClip(c);
   if(isPower) {
      Tools.setClip(c,hero_x - 9+Tools.offx, hero_y + Tools.offy - 9, 50, 50);
      c.drawBitmap(GameView.imgPower,(hero_x - 9) - nextF*50+Tools.
```

```
        offx, hero_y + Tools.offy - 9, p);
        Tools.resetClip(c);
    }
}
```

这里需要注意 action[state-1][hero_frame])%6 * 32。先看 action[state-1][hero_frame]，要结合 Hero 的各个动作的数组来解释。动作数组如下所示：

```
public int[][] action = {
        {11}, //左跳
        {6,7,8,9,10},//左
        {0,1,2,3,4},//右
        {5}, //右跳
};
```

图　8-5

数组中的每一个数字对应 Hero 的不同帧。只要改变数组的下标，就会得到 Hero 不同的动作，形成连贯的动画。在此之前定义了 Hero 的 4 种状态，如下所示：

```
public static final int STATE_LEFT  = 2;   //左走
public static final int STATE_RIGHT = 3;  //右走
public static final int STATE_JUMPL = 1;  //左跳
public static final int STATE_JUMPR = 4;  //右跳
```

二维数组 action 的第一维为 Hero 的状态，分为上述 4 种，所以只要在逻辑中改变 state 的值，便会得到 Hero 不同状态下的动作。

```
/**
    * 左移动
    */
    public void moveLeft() {
        state = STATE_LEFT;    //改变 Hero 状态
        if(isJumping) {//如果跳跃
            state = STATE_JUMPL; //改变 Hero 状态
        }
        hero_x -= hero_speed;
        if(hero_x < -4){hero_x = -4;}
        nextFrame();
```

```
    }
    /**
     * 右移动
     */
   public void moveRight() {
        state = STATE_RIGHT;    //改变 Hero 状态
        if(isJumping) {//如果跳跃
            state = STATE_JUMPR; //改变 Hero 状态
        }
        hero_x += hero_speed;
        if(hero_x + 28 > 360){hero_x = 360 - 28;}
        nextFrame();
    }
```

只要改变 action 第二维数据 hero_frame 的值，就会得到 Hero 不同状态下某个动作的帧，如下所示：

```
public void nextFrame() {
    hero_frame = hero_frame < action[state-1].length-1 ? ++hero_frame:0;
    }
```

8.2.6　道具的绘制

道具类构造如下所示：

```
    /**
 * 道具构造
 * @param bx 道具 x 坐标
 * @param by 道具 y 坐标
 * @param bKinds 道具种类
 */
public Bonus(int bx,int by,int bKinds){
    bonusX = bx;
    bonusY = by;
    bonusKind = bKinds;
}
```

绘制方法如下所示（与绘制挡板是一致的）：

```
public static void drawAllBonus(Canvas c,Bonus allBonus[],Paint p) {
    for(int i = 0;i < allBonus.length;i++) {
        if((allBonus[i].bonusX + Tools.offx + allBonus[i].BW) > 120
                && (allBonus[i].bonusX + Tools.offx) < 360
                    && (allBonus[i].bonusY + Tools.offy) < 330
                    && (allBonus[i].bonusY+Tools.offy+allBonus[i].BH)>-10){
                if(allBonus[i].isShow) {
                    Tools.setClip(c,allBonus[i].bonusX+Tools.offx,
allBonus[i].bonusY + Tools.offy, 12, 12);
```

```
                    c.drawBitmap(GameView.imgGameDaoju, allBonus[i].
                    bonusX - allBonus[i]. bonusKind* 12+Tools.offx,
                    allBonus[i].bonusY+Tools.offy,p);
                    Tools.resetClip(c);
            }
        }
    }
}
```

8.3 逻辑

8.3.1 游戏 LOGO 的逻辑

```
case LOGO:
    time++;
    if(time % 20 == 0 && logo_index == bmp_logo.length-1)
    {
        state = CHOOSE;
        cleanAllPress();
        time = 0;
    }
    if(time % 20 == 0){
        logo_index = logo_index >= (bmp_logo.length-1) ? bmp_logo.
length-1 : ++logo_index;
    }
break;
```

本游戏每秒跑 20 帧。假设有 2 张 logo 图片，则每秒过 1 张，当第 2 张 logo 结束后，state=CHOOSE;进入 CHOOSE 状态。

8.3.2 游戏菜单的逻辑

```
public void menuLogic(){
//菜单的滚动逻辑，在没有滚动完一格的时候，不释放滚动，使滚动继续
    if(isMenuRollUp){
        if(menu_count < 8){
            menuPointY -= 2;
            if(menu_count%3==0){
            menuZhuanlunY --;
            if(menuZhuanlunY < 0){
                menuZhuanlunY = 2;
            }
            }
            if(menuPointY == 0){
                menuPointY = 96;
```

```
            }
            menu_count++;
        }else{menu_count = 0;isMenuRollUp = false;}
    }
    if(isMenuRollDown){
        if(menu_count < 8){
            menuPointY += 2;
            if(menu_count%3==0){
            menuZhuanlunY ++;
            if(menuZhuanlunY > 2){
                menuZhuanlunY = 0;
            }
            }
            if(menuPointY == 112){
                menuPointY = 16;
            }
            menu_count++;
        }else{menu_count=0;isMenuRollDown = false;}
    }
}
```

游戏的菜单采用滚动式设计，也就是当按下按键后，将 isMenuRollUp 置为 true：

```
if(isMenuRollUp){
    ...............................
}
```

菜单在一定时间内从上一项滚动到下一项。例如，从“开始游戏”滚动到“继续游戏”，这个动作在 8 帧内完成。完成后，将 isMenuRollUp 置为 false，将 menu_count 置为 0。代码如下所示：

```
if(menu_count < 8){
    menuPointY -= 2;
........................
} else{menu_count = 0;isMenuRollUp = false;}
```

8.3.3 游戏背景的逻辑

详见 8.2.3 节“游戏背景绘制”。

8.3.4 游戏元素块的逻辑

根据不同的元素块类型，处理不同的逻辑，代码如下所示：

```
/**
 * 从中间伸缩
 */
public void changeByCenter(){
    if(isadd==1){
        bx -= speed;
```

```
            addx += 2*speed;
            if(addx >= board_max){
            isadd = 0;
        }
    }else{
        bx += speed;
        addx -= 2*speed;
        if(addx <= board_min){
            isadd = 1;
        }
    }
}
/**
 * 从左边伸缩
*/
public void changeByLeft(){
    if(isadd==1){
        addx -= speed;
        if(addx <= board_min){
            isadd = 0;
        }
    }else{
        addx += speed;
        if(addx >= board_max){
            isadd = 1;
        }
    }
}
/**
 * 从右边伸缩
*/
public void changeByRight(){
    if(isadd==1){
        bx -= speed;
        addx +=speed;
        if(addx >= board_max){
            isadd = 0;
        }
    }else{
        bx += speed;
        addx -= speed;
        if(addx <= board_min){
```

```
            isadd = 1;
        }
    }
}
/**
 *左右移动
 */
public void moveLeftAndRight(){
    if(bx >= board_max){speed = -speed;}
    if(bx <= board_min){speed = -speed;}
    bx += speed;
}
/**
 * 上下移动
 */
public void moveUpAndDown(){
    if(by >= board_max){speed = -speed;}
    if(by <= board_min){speed = -speed;}
    by += speed;
}
```

8.3.5 游戏人物的逻辑

```
/**
 * 左移动
 */
public void moveLeft() {
    state = STATE_LEFT;
    if(isJumping) {
        state = STATE_JUMPL;
    }
    hero_x -= hero_speed;
    if(hero_x < -4){hero_x = -4;}
    nextFrame();
}
/**
 * 右移动
 */
public void moveRight() {
    state = STATE_RIGHT;
    if(isJumping) {
        state = STATE_JUMPR;
    }
```

```
    hero_x += hero_speed;
    if(hero_x + 28 > 360){hero_x = 360 - 28;}
    nextFrame();
}

public void up() {
    if(!isDown){
        if(!isJump&&!isJumping) {
            isJump = true;
            isJumping = true;
        }
    }
}
/**
 * 播放动画
 */
public void nextFrame() {
hero_frame = hero_frame < action[state-1].length-1 ?
++hero_frame:0;
}
/**
 * 逻辑
 */
public void logic(Board[] allb) {
    if(isJump) {
        jump();
    }
    else{
        down(allb);
    }
    choiceSpeed();
    if(isPause){        //定时所有挡板与怪物 5 秒
        if(++countZ>= 100) {
            isPause = false;
            countZ = 0;
        }
    }
    if(isPower) {      //hero 无敌 5 秒
        if(++countB >= 100) {
            isPower = false;
            countB = 0;
        }
    }
    if(isAddSpeed) {  //hero 加速 8 秒
```

```
        if(++countS >= 160) {
            isAddSpeed = false;
            countS = 0;
        }
    }
    if(++nextF > 5){nextF=0;}

}
/**
 * 跳跃上升
 */
public void jump() {
    hero_frame = 0;
    if(state == STATE_LEFT) {state = STATE_JUMPL;}
    if(state == STATE_RIGHT) {state = STATE_JUMPR;}
    hero_y-= hero_jump_speed;
    hero_jump_speed --;
    if(hero_jump_speed <= 0){
        isJump = false;
    }
}
/**
 * 下降
 */
public void down(Board[] allb){
    isDown = true;
    if(hero_jump_speed < 10){
        hero_jump_speed++;
    }
    hero_y += hero_jump_speed;
    downdistance += hero_jump_speed;
    Tools.checkHeroAndBoard(this, allb); //检测是否站在了挡板上

    if(isStandBorad){ //如果是刚改变hero状态为非跳跃状态，修正hero坐标
        hero_y = onBonus - 32;
        isJumping = false;
        hero_jump_speed = 11;
        if(state == STATE_JUMPL) {state = STATE_LEFT;}
        if(state == STATE_JUMPR) {state = STATE_RIGHT;}
        isDown = false;
        if(!isPower){ //如果不在保护状态下 并且下降距离超过300，则掉一命
            if(downdistance > 300){
```

```
                life--;
                countB = 0;
                isPower = true;
            }
        }
        downdistance = 0;
    }
}
/**
 * 判断 Hero 的速度，即是否为加速状态
 */
public void choiceSpeed() {
    if(isAddSpeed) {
        hero_speed = hero_addSpeed;
    }else {
        hero_speed = 3;
    }
}
```

8.3.6　道具的逻辑

道具的逻辑就是与 Hero 的碰撞检测（矩形碰撞）。根据 Hero 碰到的不同类型的道具去改变相应的变量值，实现不同的功能，代码如下所示：

```
public static void withBonusCollision(Hero h,Bonus[] b) {
    for (int i = 0; i < b.length; i++) {
        if(b[i].bonusKind != 4 && b[i].isShow) {
            if ((h.hero_x + h.heroW) < b[i].bonusX
                    || (h.hero_y + h.heroH) < b[i].bonusY
                        || (b[i].bonusX + b[i].BW) < h.hero_x
                            || (b[i].bonusY + b[i].BH) < h.hero_y) {

            }else {
                if(b[i].bonusKind == Bonus.BS_CAP) {
                    h.countB = 0;
                    h.isPower = true;
                    b[i].isShow = false;
                }else if(b[i].bonusKind == Bonus.BS_TIGER) {
                    if(Hero.life < 8) {
                        Hero.life++;
                    }
                    b[i].isShow = false;
                }else if(b[i].bonusKind == Bonus.BS_TIMER){
                    h.countZ = 0;
                    h.isPause = true;
```

```
                b[i].isShow = false;
            }else if(b[i].bonusKind == Bonus.BS_SPEED) {
                b[i].countS = 0;
                h.isAddSpeed = true;
                b[i].isShow  = false;
            }
        }
    }
    if(b[i].bonusKind == 4) {
        if(h.isDown) {
            if (b[i].bonusY - (h.hero_y) <= 32 && b[i].bonusY -
(h.hero_y) >= 21) {
                if((b[i].bonusX < (h.hero_x + 30))
                    && ((b[i].bonusX + 12) > (h.hero_x + 12))) {
                    h.isJump = true;
                    h.isJumping = true;
                    h.downdistance = 0;
                    h.hero_jump_speed = 20;
                }
            }
        }
    }
  }
}
```

8.4 游戏按键

8.4.1 游戏菜单的按键处理

```
case MENU:
//按下一次，要等菜单条滚动完一个。所以，加上一个 isMenuRoll 变量控制
if(!isMenuRollUp){
    if(leftPress){
        isMenuRollDown = true;
    }
}
if(!isMenuRollDown){
    if(rightPress){
        isMenuRollUp = true;
    }
}
//不需要做出滚动效果，所以确认处理写在按键处理中
if(!isMenuRollUp&&!isMenuRollDown){
    if(okPress) {
        switch (menuPointY) {
```

```
            case 16://开始游戏
                startGame();
                break;
            case 32://继续游戏
                conGame();
                break;
            case 48://游戏设置
                cleanAllPress();
                OLD_STATE = MENU;
                state = SET;
                break;
            case 64://游戏帮助
                state = HELP;
                cleanAllPress();
                break;
            case 80://关于游戏
                state = ABOUT;
                cleanAllPress();
                break;
            case 96://退出游戏
                exit();
                break;
        }
    }
}
break;
```

8.4.2 游戏人物的按键处理

```
if(!isGamePause){
    if(leftPress) {
        hero.moveLeft();
    }
    if(rightPress) {
        hero.moveRight();
    }
    if(upPress || okPress) {
        hero.up();
    }
}
```

如果在非暂停状态下，根据不同的按键调用 Hero 类方法，代码如下所示：

```
if(!isGamePause){
    ……………………
}
```

8.5 附件：源代码

8.5.1 GameActivity 类

```
package com.ming.last;
import android.app.Activity;
import android.content.pm.ActivityInfo;
import android.os.Bundle;
import android.view.KeyEvent;
import android.view.Window;
import android.view.WindowManager;
public class GameActivity extends Activity {
    /** Called when the activity is first created. */
    public GameView gv;
    @Override
    public void onCreate(Bundle savedInstanceState) {
        super.onCreate(savedInstanceState);
        setRequestedOrientation(
            ActivityInfo.SCREEN_ORIENTATION_LANDSCAPE);//设置为横屏显示
        requestWindowFeature(Window.FEATURE_NO_TITLE); //设定窗口无标题栏
        getWindow().setFlags( //设置窗口为全屏
                WindowManager.LayoutParams.FLAG_FULLSCREEN,
                WindowManager.LayoutParams.FLAG_FULLSCREEN);
        gv = new GameView(this);
        setContentView(gv);//将gv设定为显示窗口
    }
    @Override
    /**
     * 重写父类方法，用于响应按键事件
     */
    public boolean onKeyDown(int keyCode, KeyEvent event) {
        if(keyCode == KeyEvent.KEYCODE_BACK && gv.isExit) {
            gv.exit();
            return super.onKeyDown(keyCode, event);
        }else {
            return gv.onKeyDown(keyCode, event);
        }
    }
}
```

8.5.2 GameView 类

GameView 类继承 SurfaceView 类，实现 Callback 及 Runnable 接口，负责控制整个游戏。

主要方法如下所示：

```
package com.ming.last;

import java.util.Random;
import android.content.Context;
import android.content.SharedPreferences;
import android.graphics.*;
import android.os.SystemClock;
import android.view.KeyEvent;
import android.view.MotionEvent;
import android.view.SurfaceHolder;
import android.view.SurfaceView;
import android.view.SurfaceHolder.Callback;
import android.widget.Toast;

public class GameView extends SurfaceView implements Callback,Runnable{
    public Tools tool;
    public Music music;
    public Hero hero;
    public Context context;
    public Bitmap bmp;
    public Bitmap bmp_logo[];
    public Bitmap
imgMenuBg,imgMenuChoose,imgMenuTiger,imgMenuZhuanlun,imgmenubk;//m
enu 中的图片
    public Bitmap
imgSetKaiguan,imgSetJiantou,imgSetMG,imgSetKuang;//set 中的图片
    public Bitmap imggameinmenu,imggameTlife,imgDeadMenu,imgGameBg;
    public Bitmap imgLoos,imgtiger,imgGameOver,imgGameLogo,imgLogoBk;
    public Bitmap imggameinMenuBg,imggameKuang,imgLoosBk,imggameover;
    public static Bitmap
imggameNpc,imgGameDaoju,imggameDoor,imgPower,imggameBoard;
    public Bitmap bmpPanel,bmpDirBtn1,bmpDirBtn2,bmpJumpBtn;
    ///////////////////////////
    public Thread t;
    public SurfaceHolder sh;

    public Paint p = new Paint();
    public int scr_w; //屏幕宽
    public int scr_h; //屏幕高
    public long startTime;
    public long endTime;
```

```
public int rate = 1000 / 20; //FPS达到15以上
public int FPS;
public boolean isPress;
public int col_x,col_y;
public byte state=0;
public final byte LOGO = 0;
public final byte MENU = 1;
public final byte GAME = 2;
public final byte CHOOSE= 3;
public final byte HELP = 4;
public final byte ABOUT = 5;
public final byte SET = 6;

////////////////////////////////////
public final byte HERO_DEAD = 8;
public final byte GAME_LOGO = 9;
public final byte GAME_OVER = 10;
public byte OLD_STATE = MENU;
///////////////////////////////////////
public int count;//控制LOGO时间的增量
public int menu_count;//控制MENU滚动的增量
public int logoIndex;//LOGO图片数组位置
public int menuPointY=16;//菜单指针坐标
public int GamePointY;//游戏指针坐标
public int menuZhuanlunY = 0;//装轮的y坐标
public int setPointX,setPointY,setPointRow;//设置菜单指针
public Random rand = new Random();//随机数
public int index; //开屏效果索引
public boolean isMenuRollUp;//菜单上滚
public boolean isMenuRollDown;//菜单下滚
public boolean isGamePause;//游戏暂停
public boolean isCanChoice;//是否可选下一关
public boolean isDrawLogo;
public Board[] allb = null;//挡板数组
public Npc[] alln = null;//怪物数组
public Bonus[] allBonus = null;//道具数组
public int level;//游戏关卡
public int beginLevel;//开始的关卡
public int levelchoice = 1;//关卡选择
public static int game_kz = 0;//开头动画控制
public static int game_sdx;//显示大小
public boolean  isReanData = true;//是否读取数据库
```

```
///////////////////按键状态
public boolean leftPress;//按下，左方向
public boolean rightPress;//按下，右方向
public boolean upPress;//按下，上方向
public boolean downPress;//按下，下方向
public boolean left_softPress;//按下，左软键
public boolean right_softPress;//按下，右软键
public boolean okPress;//按下，OK 键
public boolean anyKeyPress;//按下，任意键
public final int BUTTON_OFFX = 55;
public int left_btn_offx;
public int right_btn_offx;
public int up_btn_offx;
public int down_btn_offx;
public int ok_btn_offx;
////////////////////////////////////
////////////临时的生命　关卡　最大关卡
public int lsLife,lsLevel,lsMaxLevel;
public int time = 0;
public int logo_index = 0;
public boolean isRun;
public boolean isExit;
public int exitCon;
public int con;
///////////////////////////////////
//                    1 2 3 4 5 6 7 8 9 10 11 12 13 14 15
public int levelArray[] = {1,0,8,6,14,7,5,3,9,2,4,10,11,13,12};
public int setArray[][] = {
       {234,140,87,20},//{x 坐标,y 坐标,图宽,图高/2}箭头图
       {160,143,55,13},//{x 坐标,y 坐标,图宽/2,图高}文字图
       {222,140},//{x 坐标,y 坐标}选择框图
       {253,140,44,19},//{x 坐标,y 坐标,图宽/2,图高}开关图
       {37},//{行距}
       {257,193}//关卡
};
public int maxLevel; //可选择的最大关卡
public int point;    //失败菜单指针
public int setCount; //菜单背景播放计数器
public static int tmph;     //英雄样式随机
public int tmpb;     //背景样式随机
public static int tmpT;     //挡板样式随机
public int frame;    //hero 死亡帧
```

```
public int deadCon;  //控制hero死亡帧的播放时间
public int lastlife; //最后的生命数
private boolean isBeginNewLevel;//是否开始新的一关
private int countNextLevel;//控制显示关卡的时间
private int nextPage;
////////////////////////////////////////
public final int CHOOSE_TEXT_SIZE = 20;
public Rect dir_rect[]  = new Rect[2];
public Rect ok,leftSoft,rightSoft,closeMusic,startMusic;
public final int  CHOOSE_TEXT_COLOR = Color.WHITE,
              CHOOSE_TEXT_TOUCH_COLOR = Color.YELLOW;
public int choose_color = CHOOSE_TEXT_COLOR;

////////////////////////////////////////
public int touchX,touchY;
/////////////////////////

public GameView(Context context) {
    super(context);
    this.context = context;
    tool = new Tools(context);  //构造Tools类
    music = new Music(context); //构造Music类
    loadImage();                //加载图片资源
    hero = new Hero();          //构造Hero类
    p.setAntiAlias(true);       //削除所画图形边缘的锯齿
    t = new Thread(this);       //创建线程
    sh = getHolder();          //获取Holder
    sh.addCallback(this);       //添加回调
    setFocusable(true);         //添加触摸焦点
}
public void loadImage() {
    imgGameLogo     = tool.createBmp(R.drawable.b_gamelogo);
    imgLogoBk       = tool.createBmp(R.drawable.b_game_logobk);
    imgmenubk       = tool.createBmp(R.drawable.c_menubk);
    imgMenuBg       = tool.createBmp(R.drawable.c_menubg);
    imgMenuChoose   = tool.createBmp(R.drawable.c_menuchoose);
    imgMenuTiger    = tool.createBmp(R.drawable.c_menutiger);
    imgMenuZhuanlun = tool.createBmp(R.drawable.c_menuzhuanlun);
    imgSetKaiguan   = tool.createBmp(R.drawable.c_setkaiguan);
    imgSetKuang     = tool.createBmp(R.drawable.c_setkuang);
    imgSetMG        = tool.createBmp(R.drawable.c_setmg);
    imgSetJiantou   = tool.createBmp(R.drawable.c_setjiantou);
```

```
    imgGameDaoju    = tool.createBmp(R.drawable.d_gamedaoju1);

    bmpPanel        = tool.createBmp(R.drawable.panel);
    bmpDirBtn1       = tool.createBmp(R.drawable.dirbtn1);
    bmpDirBtn2       = tool.createBmp(R.drawable.dirbtn2);
    bmpJumpBtn      = tool.createBmp(R.drawable.jbtn);

    imggameBoard    = tool.createBmp(R.drawable.d_board);
    imggameTlife    = tool.createBmp(R.drawable.d_gamelife);
    imggameKuang    = tool.createBmp(R.drawable.d_gamekuang);
    imggameinmenu   = tool.createBmp(R.drawable.d_gameinzi);
    imggameinMenuBg = tool.createBmp(R.drawable.d_gameinbenubg);
    imgLoosBk       = tool.createBmp(R.drawable.d_loosbk);
    imgLoos         = tool.createBmp(R.drawable.d_loos);
    imgGameOver     = tool.createBmp(R.drawable.d_gameover);
    imgtiger        = tool.createBmp(R.drawable.d_tiger);
    imgGameBg       = tool.createBmp(R.drawable.d_gamebk);
    imggameDoor     = tool.createBmp(R.drawable.d_gamedoor);
    imggameNpc      = tool.createBmp(R.drawable.d_gamenpc);
    imgPower        = tool.createBmp(R.drawable.d_gameshield);
    imggameover     = tool.createBmp(R.drawable.d_gamehappy);
}
public void createRand() {
    tmph = Math.abs(rand.nextInt())%3;
    tmpb = Math.abs(rand.nextInt())%3;
    tmpT = Math.abs(rand.nextInt())%3;
}
public void init(){
    bmp_logo = new Bitmap[2];
    final int[] bmp_array = {0x7f020000};
    bmp_logo = tool.createBmp(bmp_array, bmp_array.length);
    scr_w = this.getWidth();
    scr_h = this.getHeight();
    //方向键初始化
    dir_rect[0] = touchRect(0,170,60,55);   //left
    dir_rect[1] = touchRect(65,170,60,55);  //right
    /*************************************************************/
    //左、右软键初始化
    leftSoft = touchRect(0, scr_h-30, 60, 30);
    rightSoft = touchRect(scr_w-60, scr_h-30, 60, 30);
/*************************************************************/
    //game 键初始化
```

```
        ok = touchRect(400, 190, 64, 64);
    }
    /**
     * 设定矩形触碰框
     * @param x
     * @param y
     * @param w
     * @param h
     * @return  返回矩形
     */
    public Rect touchRect(int x,int y,int w,int h) {
        return new Rect(x,y,x+w,y+h);
    }
    /**
     * 清除 logo 所用图片
     */
    public void clean() {
        if(bmp_logo != null) {
            bmp_logo = null;
        }
    }
    /**
     * 清除前一关的所有对象数组
     */
    public void cleanAll(){
        allb = null;
        alln = null;
        allBonus = null;
    }
    /**
     * 重载关卡，初始化每一关的所有数据
     */
    public void initGame(){
        cleanAll();
        createRand();

    hero.initHero(imgtiger,Map.hero_array[levelArray[level]][0],Map
.hero_array[levelArray[level]][1],Map.hero_array[levelArray[level]]
[2]);
        allb = new Board[Map.game_map[levelArray[level]].length];
        if(allb.length > 0) {
            for (int i = 0; i < allb.length; i++) {
```

```
                allb[i] = new
Board(Map.game_map[levelArray[level]][i][0],Map.game_map[levelArray
[level]][i][1],

    Map.game_map[levelArray[level]][i][2],Map.game_map[levelArray[level]]
[i][3],Map.game_map[levelArray[level]][i][4],

    Map.game_map[levelArray[level]][i][5],Map.game_map[levelArray[level]]
[i][6],Map.game_map[levelArray[level]][i][7]);
            }
        }
        alln = new Npc[Map.npc_array[levelArray[level]].length];
        if(alln.length > 0) {
            for (int i = 0; i < alln.length; i++) {
                alln[i] = new
Npc(Map.npc_array[levelArray[level]][i][0],Map.npc_array[levelArra
y[level]][i][1],

    Map.npc_array[levelArray[level]][i][2],Map.npc_array[levelArray[level]]
[i][3],

    Map.npc_array[levelArray[level]][i][4],Map.npc_array[levelArray[level]]
[i][5]);
            }
        }
        allBonus = new Bonus[Map.bonus[levelArray[level]].length];
        if(allBonus.length > 0) {
            for (int i = 0; i < allBonus.length; i++) {
                allBonus[i] = new
Bonus(Map.bonus[levelArray[level]][i][0],Map.bonus[levelArray[level]]
[i][1],
                        Map.bonus[levelArray[level]][i][2]);
            }
        }
        index = Math.abs(rand.nextInt())%5;
        lastlife = Hero.life;
    }
    /**
     * 开始游戏
     */
    public void startGame() {
        level = beginLevel;
```

```
    beginLevel = 0;
    Hero.life = 3;
    game_kz = 0;
    isBeginNewLevel = true;
    isReanData = false;
    initGame();
    state = GAME;
    cleanAllPress();
}
/**
 * 继续游戏
 */
public void conGame() {
    if(Hero.life < 0) {
        return;
    }else {
        if(OLD_STATE==MENU&&isReanData){
            restoreStage();
            Hero.life = lsLife;
            level = lsLevel;
            isBeginNewLevel = true;
            initGame();
        }
        state = GAME;
        cleanAllPress();
    }
}
/**
 *
 */
public void cleanAllPress(){
        leftPress = false;
        rightPress = false;
        upPress = false;
        downPress = false;
        left_softPress = false;
        right_softPress = false;
        okPress = false;
        anyKeyPress = false;
}

/**
```

```
 * 检测四个方向键是否被按下
 * @param tx 触点 x
 * @param ty 触点 y
 * @param upOrDown 如果检测按下，传 true；如果检测抬起，传 false
 */
public void checkDirButtonEvent(int tx,int ty,boolean upOrDown) {
    for(int i = 0;i < dir_rect.length;i++) {
        if(tool.checkRectTouch(tx, ty,dir_rect[i])) {
            switch(i) {
            case 0:
                leftPress = upOrDown;
                break;
            case 1:
                rightPress = upOrDown;
                break;
            }
        }
    }
}
@Override
/**
 * 触屏处理
 */
public boolean onTouchEvent(MotionEvent me) {
    int tx = (int)me.getX();
    int ty = (int)me.getY();
    switch(me.getAction()) {
    case MotionEvent.ACTION_DOWN:  //按下处理
    switch(state) { //根据不同的状态做不同的处理
        case HELP:
        if(tool.checkRectTouch(tx, ty, leftSoft)) {
            left_softPress = true;
            if(nextPage  == -320) {
                    nextPage = 0;
                    left_softPress = false;
                }else if(nextPage == 0){
                    nextPage = -320;
                    left_softPress = false;
                }
            }
        if(tool.checkRectTouch(tx, ty, rightSoft)) {
                right_softPress = true;
```

```
                nextPage = 0;
                state = MENU;
                cleanAllPress();
            }
            break;
        case ABOUT:
            if(tool.checkRectTouch(tx, ty, rightSoft)) {
                right_softPress = true;
            }
            break;
        case SET:
            if(tool.checkRectTouch(tx, ty, dir_rect[0])){
                    if(setPointRow==0){
                    setPointX --;
                    if(setPointX < 0){setPointX = 1;}
                }else{
                    levelchoice --;
                    if(levelchoice < 1){levelchoice = 1;}
                }
                leftPress = true;
            }
            if(tool.checkRectTouch(tx, ty, dir_rect[1])){
                if(setPointRow==0){
                    setPointX ++;
                    if(setPointX > 1){setPointX = 0;}
                }else{
                    if((levelchoice - 1) < maxLevel){
                        levelchoice++;
                        if(levelchoice > 15){levelchoice = 15;}
                    }else{
                        isCanChoice = true;
                    }
                }
                rightPress = true;
            }
            if(tool.checkRectTouch(tx, ty, rightSoft)) {
                right_softPress = true;
                levelchoice = 1;
                state = OLD_STATE;
                cleanAllPress();
            }
            if(OLD_STATE != GAME){
```

```
                if(tool.checkRectTouch(tx, ty, leftSoft)) {
                left_softPress = true;
                beginLevel = levelchoice - 1;
                    levelchoice = 1;
                    state = OLD_STATE;
                    cleanAllPress();
                }
            }
                if(tool.checkRectTouch(tx, ty, ok)) {
                okPress = true;
                setPointRow ^= 1;
            }
            break;
        case GAME:
            if(!isBeginNewLevel && game_kz == -1) {
                if(tool.checkRectTouch(tx, ty,rightSoft)){
                    right_softPress = true;
                    isGamePause = !isGamePause;
                }
            }
            if(!isGamePause){
                checkDirButtonEvent(tx,ty,true);
                if(tool.checkRectTouch(tx, ty, ok)) {
                    okPress = true;
                }
            }else{
                if(tool.checkRectTouch(tx, ty, dir_rect[0])){
                    leftPress = true;
                    GamePointY -= 1;
                    if(GamePointY < 0){
                        GamePointY = 3;
                    }
                }
                if(tool.checkRectTouch(tx, ty, dir_rect[1])){
                    rightPress = true;
                    GamePointY += 1;
                    if(GamePointY > 3){
                        GamePointY = 0;
                    }
                }
                if(tool.checkRectTouch(tx, ty, ok)){
                    okPress = true;
```

```
                switch (GamePointY) {
                case 0://继续游戏
                    isGamePause = false;
                    break;
                case 1://声音设置
                    GamePointY = 0;
                    isGamePause = false;
                    OLD_STATE = GAME;
                    state = SET;
                    cleanAllPress();
                    break;
                case 2://重新开始本关
                    GamePointY = 0;
                    isGamePause = false;
                    Hero.life = lastlife;
                    initGame();
                    break;
                case 3://返回菜单
                    GamePointY = 0;
                    state = MENU;
                    OLD_STATE = GAME;
                    isGamePause = false;
                    cleanAllPress();
                    break;
                }
            }
        }
        break;
    case MENU:
        checkDirButtonEvent(tx,ty,true);
        if(tool.checkRectTouch(tx, ty,
ok)||tool.checkRectTouch(tx, ty, leftSoft)) {
            okPress = true;
        left_softPress = true;
        }
        break;
    case CHOOSE:
        if(tool.checkRectTouch(tx, ty, touchRect(0,
scr_h-CHOOSE_TEXT_SIZE*2-10, CHOOSE_TEXT_SIZE+10,
CHOOSE_TEXT_SIZE+10))) {
            left_softPress = true;
        }
```

```
            if(tool.checkRectTouch(tx,   ty,   touchRect(scr_w  -
CHOOSE_TEXT_SIZE-10, scr_h-CHOOSE_TEXT_SIZE*2-10, CHOOSE_TEXT_SIZE+10,
CHOOSE_TEXT_SIZE+10))) {
                right_softPress = true;
            }
            break;
        case LOGO:
            if(tool.checkRectTouch(tx, ty, touchRect(0, 0, scr_w,
scr_h))) {
                anyKeyPress = true;
            }
            break;
        case GAME_OVER:
            if(tool.checkRectTouch(tx, ty, rightSoft)) {
                right_softPress = true;
            }
            break;
        case HERO_DEAD:
            if(tool.checkRectTouch(tx, ty, dir_rect[0])) {
                leftPress = true;
                if(--point < 0) {
                    point = 1;
                }
            }
            if(tool.checkRectTouch(tx, ty, dir_rect[1])) {
                rightPress = true;
                if(++point > 1) {
                    point = 0;
                }
            }
            if(tool.checkRectTouch(tx, ty,
ok)||tool.checkRectTouch(tx, ty, leftSoft)) {
                okPress = true;
                left_softPress = true;
                switch (point) {
                case 0:
                    Hero.life = 3;
                    initGame();
                    state = GAME;
                    cleanAllPress();
                    break;
                case 1:
```

```
                    state = MENU;
                    cleanAllPress();
                    break;
                }
            }
            break;
        }
    break;

        case MotionEvent.ACTION_UP:  //抬起处理
        switch(state) {
            case HELP:
            if(tool.checkRectTouch(tx, ty, leftSoft)) {
                left_softPress = false;
            }
            break;
        case ABOUT:
            break;
        case SET:
            checkDirButtonEvent(tx, ty, false);
            if(tool.checkRectTouch(tx, ty, ok)) {
                okPress = false;
            }
            if(tool.checkRectTouch(tx, ty,rightSoft)){
                right_softPress = false;
            }
            if(tool.checkRectTouch(tx, ty, leftSoft)) {
                left_softPress = false;
            }
            break;
        case GAME:
            checkDirButtonEvent(tx, ty, false);
            if(tool.checkRectTouch(tx, ty, ok)) {
                okPress = false;
            }
            if(tool.checkRectTouch(tx, ty,rightSoft)){
                right_softPress = false;
            }
            break;
        case MENU:
            checkDirButtonEvent(tx, ty, false);
            if(tool.checkRectTouch(tx, ty, ok)) {
```

```
            okPress = false;
        }
        if(tool.checkRectTouch(tx, ty, leftSoft)) {
            left_softPress = false;
        }
        break;
    }
    break;
    case MotionEvent.ACTION_MOVE: //滑动处理
    switch (state) {
    case GAME:
        checkDirButtonEventMove(tx, ty, false);
        if(!tool.checkRectTouch(tx, ty, ok)) {
            okPress = false;
        }
        if(!tool.checkRectTouch(tx, ty,rightSoft)){
            right_softPress = false;
        }
        if((touchY > ty)&&(touchX > tx)){
            rightPress = true;
            okPress = true;
        }
        break;
    case MENU:
        checkDirButtonEventMove(tx, ty, false);
        if(!tool.checkRectTouch(tx, ty, ok)) {
            okPress = false;
        }
        if(!tool.checkRectTouch(tx, ty, leftSoft)) {
            left_softPress = false;
        }
        break;
    case SET:
        checkDirButtonEventMove(tx, ty, false);
        if(!tool.checkRectTouch(tx, ty, ok)) {
            okPress = false;
        }
        if(!tool.checkRectTouch(tx, ty,rightSoft)){
            right_softPress = false;
        }
        if(!tool.checkRectTouch(tx, ty, leftSoft)) {
            left_softPress = false;
```

```
                }
                break;
                }
            break;
        }
        return true;
    }
    /**
     * 检测四个方向键是否被按下
     * @param tx 触点 x
     * @param ty 触点 y
     * @param upOrDown 如果检测按下，传 true；如果检测抬起，传 false
     */
  public void checkDirButtonEventMove(int tx,int ty,boolean upOrDown)
{
        for(int i = 0;i < dir_rect.length;i++) {
            if(!tool.checkRectTouch(tx, ty,dir_rect[i])) {
                switch(i) {
                case 0:
                    leftPress = upOrDown;
                    break;
                case 1:
                    rightPress = upOrDown;
                    break;
                }
            }
        }
    }
    @Override
    /**
     * SurfaceView 发生改变时，系统调用此方法
     */
    public void surfaceChanged(SurfaceHolder holder, int format, int
width,
            int height) {
    }
    @Override
    /**
     * SurfaceView 创建时，系统调用此方法
     */
    public void surfaceCreated(SurfaceHolder holder) {
        init();
```

```
        isRun = true;
        t.start();  //启动线程
    }

    @Override
    /**
     * SurfaceView 销毁时，系统调用此方法
     */
    public void surfaceDestroyed(SurfaceHolder holder) {
        music.stop();
        this.saveStage(Hero.life, level, maxLevel);
        isRun = false;
    }

    @Override
    public boolean onKeyDown(int keyCode, KeyEvent event) {
        if(keyCode == KeyEvent.KEYCODE_BACK ){
            con ++;
            if(con <= 1) {
                Toast.makeText(context, "再次按下退出", 0).show();
                isExit = true;
            }
        }
        return super.onKeyDown(keyCode, event);
    }
    @Override
    public boolean onKeyUp(int keyCode, KeyEvent event) {
        return super.onKeyUp(keyCode, event);
    }
    public void exit() {
        saveStage(Hero.life, level, maxLevel);
        android.os.Process.killProcess(android.os.Process.myPid());
    }
    public void logic(){
        if(isExit) {
            if(++exitCon >= 40) {
                isExit = false;
                exitCon = 0;
                con = 0;
            }
        }
        switch(state){
```

```
        case LOGO:
            time++;
            if(time % 20 == 0 && logo_index == bmp_logo.length-1)
            {
                state = CHOOSE;
                cleanAllPress();
                time = 0;
            }
            if(time % 20 == 0){
                logo_index  =  logo_index  >=  (bmp_logo.length-1)  ?
bmp_logo.length-1 : ++logo_index;
            }
            break;
        case MENU:
            buttonLogic();
            menuLogic();
            break;
        case GAME:
            buttonLogic();
            if(!isGamePause){
                gameLogic();
            }
            break;
        case CHOOSE:
            break;
        case SET:
            setLogic();
            break;
        case HELP:
            break;
        case ABOUT:
            break;
        case HERO_DEAD:
            buttonLogic();
            if(++deadCon>=5) {
                frame = frame < 3 ? ++frame : 0;
                deadCon = 0;
            }
            break;
        }
    }
    /**
```

```
     * 游戏逻辑
     */
    public void gameLogic() {
        if(isBeginNewLevel){
            if(++countNextLevel>40){
                countNextLevel=0;
                isBeginNewLevel = false;
            }
        }else{
            Tools.withBonusCollision(hero, allBonus);
            if(!hero.isPause){
                Tools.moveAllBoard(allb);
                Tools.moveAllNpc(alln, allb);
            }
            Tools.checkHeroAndNpc(hero, alln);
            hero.logic(allb);
            Tools.setCamera(hero);
            checkNextLevel(Map.guanqia[levelArray[level]][0],
Map.guanqia[levelArray[level]][1], Map.guanqia[levelArray[level]][2],
hero);
            if(Hero.life < 0) {
                setPointX = 1;
                state = HERO_DEAD;
                cleanAllPress();
            }
        }
        if(!leftPress&&!rightPress&&!hero.isJump&&!hero.isJumping){
            hero.hero_frame = 2;
        }
    }
    /**
     * SET 逻辑
     */
    public void setLogic(){
        if(setPointRow == 0) {
            if(setPointX == 0) {music.start();}
            if(setPointX == 1) {music.pause();}
        }
        if(isCanChoice) {
            if(++setCount > 5){
                isCanChoice = false;
                setCount = 0;
```

```
        }
    }
}
public void buttonLogic() {
    if(leftPress) {
        left_btn_offx = BUTTON_OFFX;
    }else {
        left_btn_offx = 0;
    }
    if(rightPress) {
        right_btn_offx = BUTTON_OFFX;
    }else {
        right_btn_offx = 0;
    }
    if(okPress) {
        ok_btn_offx = 64;
    }else {
        ok_btn_offx = 0;
    }
}
/**
 * 过关判定
 * @param x
 * @param y
 * @param x1
 * @param y1
 * @param h
 */
public void checkNextLevel(int x ,int yb , int ys , Hero h){
    if (h.hero_x <= x && (h.hero_x + h.heroW) >= x &&
            (h.hero_y + h.heroH) <= ys && (h.hero_y + h.heroH) >= (ys
        - h.heroH) ) {
        level++;
        if(level <= 14){
            if (level > maxLevel) {
                maxLevel = level;
            }
            game_kz = 0;
            isBeginNewLevel = true;
            initGame();
        }else{
            level--;
```

```
                state = GAME_OVER;
                cleanAllPress();
                h.initHero(imgtiger,
Map.hero_array[levelArray[level]][0],
Map.hero_array[levelArray[level]][1],
Map.hero_array[levelArray[level]][2]);
                Hero.life = 3;
                isBeginNewLevel = true;
                game_kz = 0;
            }
        }
    }
/**
 * 绘制 logo (图 8-6)
*/
public void drawLogo(Canvas c){
    c.drawColor(Color.WHITE);
    c.drawBitmap(bmp_logo[logo_index], (scr_w-240)>>1,
            (scr_h-320)>>1, p);
}
```

图 8-6

```
/**
* 绘制音乐选择 (图 8-7)
*/

public void drawChoose(Canvas c) {
        c.drawColor(Color.BLACK);
        p.setTextSize(CHOOSE_TEXT_SIZE);
        p.setColor(Color.WHITE);
```

```
        c.drawText("是否开启音乐", (scr_w >> 1)-CHOOSE_TEXT_SIZE * 3,
         scr_h >> 1, p);
        if(left_softPress) {
            choose_color = CHOOSE_TEXT_TOUCH_COLOR;
        }else {
            choose_color = CHOOSE_TEXT_COLOR;
        }
        p.setColor(choose_color);
        c.drawText("是",10 , scr_h - CHOOSE_TEXT_SIZE - 10,p);
        c.drawText("否", scr_w - CHOOSE_TEXT_SIZE - 10, scr_h -
        CHOOSE_TEXT_SIZE-10, p);
}
```

图 8-7

```
/**
 * 绘制菜单（图 8-8）
 */
public void drawMenu(Canvas c){
    c.drawBitmap(imgmenubk, 0 + 120, 0, p);
    Tools.setClip(c,53 + 120, 260, 42, 17);
    c.drawBitmap(imgMenuZhuanlun, 53 + 120, 260 - menuZhuanlunY*17, p);
    Tools.resetClip(c);
    c.drawBitmap(imgMenuBg, 93 + 120, 256, p);
        Tools.setClip(c,103 + 120, 262, 62, 16);
        c.drawBitmap(imgMenuChoose, 103 + 120, 262 - menuPointY, p);
        Tools.resetClip(c);
        c.drawBitmap(imgMenuTiger, 30 + 120, 235, p);
        drawButton(c);
        drawSoft(c,"确定","");
    }
```

图　8-8

```
/**
 * 绘制游戏画面（图 8-9）
 * @param g
 */
public void drawGame(Canvas g) {
    if(isBeginNewLevel){
        int lev = level+1;
        g.drawColor(Color.BLACK);
        g.drawText("第"+lev+"关", (scr_w>>1) - 30, scr_h>>1, p);
    }else{
        Tools.setClip(g, 120, 0, 240, 320);
        g.drawBitmap(imgGameBg, 0-tmpb*360+Tools.offx, Tools.offy,
p);
        Tools.resetClip(g);
        Tools.drawAllBoard(g,allb,p);
        Tools.drawAllNpc(g, alln,p);
        Tools.drawAllBonus(g, allBonus,p);
        g.drawBitmap(imggameDoor,
Map.guanqia[levelArray[level]][0]-20+Tools.offx,
Map.guanqia[levelArray[level]][1] + Tools.offy, p);
        hero.paint(g,p);
        g.drawBitmap(imggameTlife, 120, 0, p);
        g.drawText(" X "+Hero.life,142,15,p);
        if(hero.isPower) {
            g.drawText("保护: "+(5 - hero.countB/20),80 + 120,130,p);
        }
```

```
        if(hero.isAddSpeed) {
            g.drawText("加速: "+(8-hero.countS/20),80+120,150,p);
        }
        if(isGamePause) {
            g.drawBitmap(imggameinMenuBg, 78 + 120, 101, p);
            g.drawBitmap(imggameKuang, 82+120, 130+GamePointY*18,
p);
            g.drawBitmap(imggameinmenu, 96 + 120, 135, p);
        }
        /*******************************************/
        drawButton(g);
            p.setColor(Color.BLACK);
            openScreen(g, index);
        drawSoft(g,"","暂停");
    }
}
```

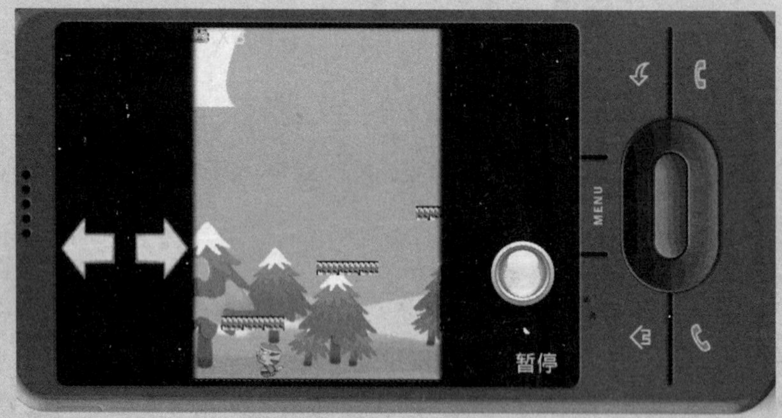

图 8-9

```
/**
 * 绘制左右软按件和菜单（图 8-10）
 * @param c
```

图 8-10

```
 */
public void drawSoft(Canvas c,String left,String right) {
    p.setColor(Color.GREEN);
    p.setTextSize(20);
    c.drawText(left, 15, scr_h-10, p);
    c.drawText(right, scr_w-40-15, scr_h-10, p);
}
/**
 * 绘制按键
 * @param c
 */
public void drawButton(Canvas c) {
    c.drawBitmap(bmpPanel, 0,0, p);
    Tools.setClip(c, 0, 170, 55, 60);
    c.drawBitmap(bmpDirBtn1,0 - left_btn_offx ,170, p);
    Tools.resetClip(c);
    Tools.setClip(c, 65, 170, 55, 60);
    c.drawBitmap(bmpDirBtn2, 65 - right_btn_offx, 170, p);
    Tools.resetClip(c);
    Tools.setClip(c, 400, 190, 64, 64);
    c.drawBitmap(bmpJumpBtn, 400 - ok_btn_offx, 190, p);
    Tools.resetClip(c);
}
/**
 * 绘制失败画面（图8-11）
 * @param g
 */
public void drawLoos(Canvas c) {
    c.drawBitmap(imgLoosBk, 120, 0, p);
    c.drawBitmap(imgLoos, 181, 50, p);
    c.drawBitmap(imggameinMenuBg, 199, 130, p);
    c.drawBitmap(imggameKuang, 202, 175+point*18, p);
    Tools.setClip(c, 216, 180, 48, 34);
    c.drawBitmap(imggameinmenu, 216,144,p);
    Tools.resetClip(c);
    Tools.setClip(c, 224, 148, 32, 32);
    c.drawBitmap(imgGameOver, 224 - frame * 32, 145-tmph*32, p);
    Tools.resetClip(c);
    drawButton(c);
    drawSoft(c,"确定","");
}
```

图 8-11

```
/**
 * 绘制“帮助”页面（图 8-12）
 */
public void drawHelp(Canvas g){
    g.drawColor(Color.BLACK);
    for (int i = 0; i < Tools.strHelp.length; i++) {
        switch (i) {
        case 15:
            Tools.setClip(g,140, i*15+nextPage, 12, 12);
            g.drawBitmap(imgGameDaoju, 140, i*15+nextPage, p);
            Tools.resetClip(g);
            break;
        case 19:
            Tools.setClip(g,140, i*15+nextPage, 12, 12);
            g.drawBitmap(imgGameDaoju, 140-12, i*15+nextPage, p);
            Tools.resetClip(g);
            break;
        case 25:
            Tools.setClip(g,140, i*15+nextPage, 12, 12);
            g.drawBitmap(imgGameDaoju, 140-24, i*15+5+nextPage, p);
            Tools.resetClip(g);
            break;
        case 31:
            Tools.setClip(g,140, i*15+nextPage, 12, 12);
            g.drawBitmap(imgGameDaoju, 140-36, i*15+nextPage, p);
            Tools.resetClip(g);
            break;
        case 35:
            Tools.setClip(g,140, i*15+nextPage, 12, 12);
            g.drawBitmap(imgGameDaoju, 140-48, i*15+nextPage, p);
            Tools.resetClip(g);
```

```
                break;
            }

        g.drawText(Tools.strHelp[i],155, 30 + i * 15+nextPage, p);
        }
        p.setColor(Color.WHITE);
        drawSoft(g,"翻页","返回");
    }
```

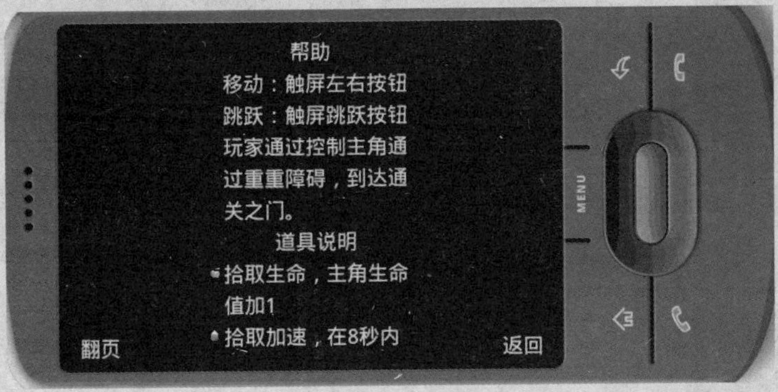

图　8-12

```
/**
 * 绘制“自我介绍”页面（图 8-13）
 */

    public void drawAbout(Canvas g) {
        g.drawColor(Color.BLACK);
        for (int i = 0; i < Tools.strAbout.length; i++) {
            g.drawText(Tools.strAbout[i],155, 30 + i * 15, p);
        }
        drawSoft(g,"","返回");
    }
```

图　8-13

```
/**
 * 绘制“设置”页面（图 8-14）
```

```
    */
   public void drawSet(Canvas c){
       c.drawColor(Color.WHITE);
       c.drawBitmap(imgSetKuang,  setArray[2][0],  setArray[2][1]  +
setPointRow*setArray[4][0], p); //选择框
       if(setPointRow==0){
           Tools.setClip(c,setArray[0][0], setArray[0][1],
setArray[0][2], setArray[0][3]);
           c.drawBitmap(imgSetJiantou, setArray[0][0], setArray[0][1] -
setArray[0][3], p);
           Tools.resetClip(c);
           Tools.setClip(c,setArray[0][0], setArray[0][1] +
setArray[4][0], setArray[0][2], setArray[0][3]);
           c.drawBitmap(imgSetJiantou, setArray[0][0], setArray[0][1] +
setArray[4][0] , p);
           Tools.resetClip(c);
       }else{
           Tools.setClip(c,setArray[0][0], setArray[0][1],
setArray[0][2], setArray[0][3]);
           c.drawBitmap(imgSetJiantou, setArray[0][0], setArray[0][1],
p);
           Tools.resetClip(c);
           Tools.setClip(c,setArray[0][0], setArray[0][1] +
setArray[4][0], setArray[0][2], setArray[0][3]);
           c.drawBitmap(imgSetJiantou, setArray[0][0], setArray[0][1] +
setArray[4][0] - setArray[0][3], p);
           Tools.resetClip(c);
       }
       //文字图
       Tools.setClip(c,setArray[1][0], setArray[1][1], setArray[1][2],
setArray[1][3]);
       c.drawBitmap(imgSetMG, setArray[1][0], setArray[1][1],p);
       Tools.resetClip(c);
       Tools.setClip(c,setArray[1][0], setArray[1][1] + setArray[4][0],
setArray[1][2], setArray[1][3]);
       c.drawBitmap(imgSetMG,   setArray[1][0]   -   setArray[1][2],
setArray[1][1] + setArray[4][0],p);
       Tools.resetClip(c);
       ///////////////////////////
       //开关
       Tools.setClip(c,setArray[3][0], setArray[3][1], setArray[3][2],
setArray[3][3]);
```

```
        c.drawBitmap(imgSetKaiguan, setArray[3][0] -
setPointX*setArray[3][2], setArray[3][1], p);
        Tools.resetClip(c);
        //关卡选择
        p.setColor(Color.BLACK);
        p.setTextSize(16);
        c.drawText(" 第 " + levelchoice + " 关 ", setArray[5][0],
setArray[5][1],p);
        drawButton(c);
        if(OLD_STATE != GAME) {
            drawSoft(c,"确定","返回");
        }else{
            drawSoft(c,"","返回");
        }
        if(isCanChoice){
            if(levelchoice < 15) {
                c.drawText("关卡未被激活", (Tools.screenW>>1) + 60, 260,
p);
            }else {
                c.drawText("已是最后一关", (Tools.screenW>>1) + 60, 260,
p);
            }
        }
    }
```

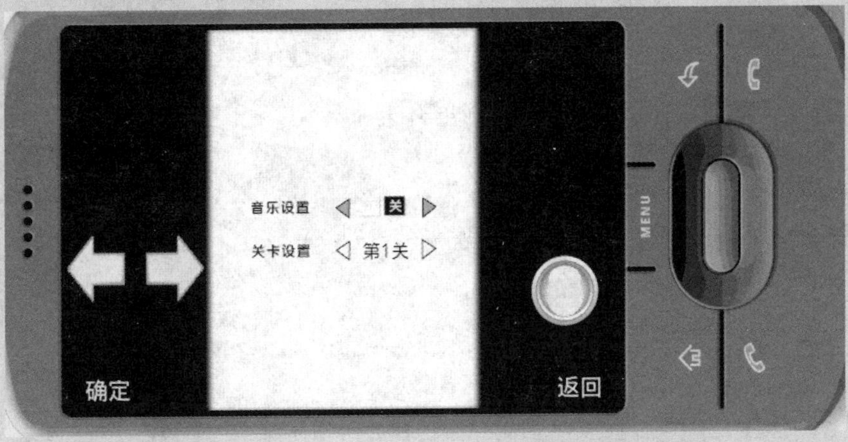

图 8-14

```
    /**
     * 绘制“结束”页面（图 8-15）
     */
    public void drawGameOver(Canvas c){
        c.drawColor(Color.WHITE);
        c.drawBitmap(imgLogoBk, 120, 0, p);
```

```
    for (int i = 4; i >= 0; i--) {
        Tools.setClip(c,50 + 120, i*65, 100, 70);
        c.drawBitmap(imgGameLogo, 50 - i*100 + 120, i*65, p);
        Tools.resetClip(c);
    }
    c.drawBitmap(imggameover, 120, 0, p);
    drawSoft(c,"","返回");
}
```

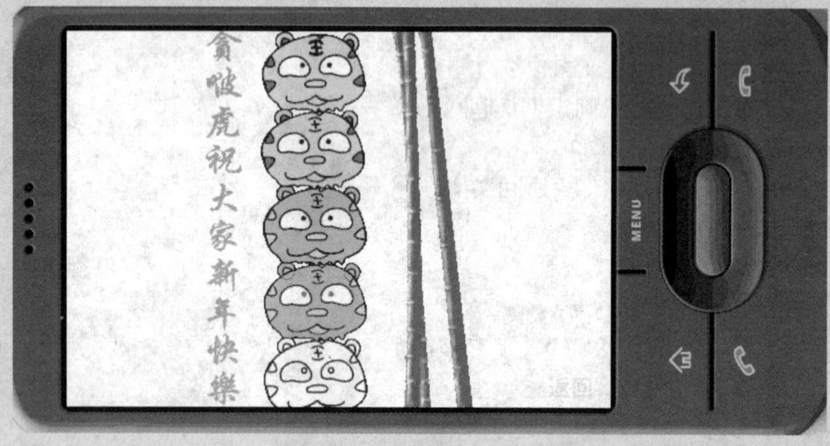

图 8-15

```
/**
 *各种开窗效果（图 8-16）
 */
public void openScreen(Canvas c , int index) {
    if(game_kz == -1)
        return;
    else if(index > -1 && index<4){
        game_sdx = game_kz;
    }
    else if(index > 3 && index < 5){
        game_sdx = game_kz*6;
    }
    if(index ==0){
        for(int i=0;i<10;i++){
                fillRect(c,120,i*32,240,32-game_sdx);
        }
    }
    else if(index == 1){
        for(int i=0;i<10;i++){
                fillRect(c,i*24 + 120,0,24 - game_sdx,scr_h);
        }
    }
```

```
        else if(index == 2){
            for(int i=0;i<6;i++){
                for(int j=0;j<8;j++) {
                    fillRect(c,i*40 + game_sdx + 120 ,j*40 + game_sdx,40 - 2*game_sdx,40 - 2*game_sdx);
                }
            }
        }
        else if(index == 3){
            for(int i=0;i<6;i++){
                for(int j=0;j<8;j++){
                    c.drawCircle(i*40 + 140,j*40 + 20,20 - 2*game_sdx,p);
                }
            }
        }
        else if(index == 4){
            fillRect(c,0+120, 0, 120-((game_sdx/3)*2),160-game_sdx);
            fillRect(c,120  +  ((game_sdx/3)*2)  +  120,  0,  120  - ((game_sdx/3)*2), 160 - game_sdx);
            fillRect(c,0 + 120, 160 + game_sdx, 120 - ((game_sdx/3)*2), 160 - game_sdx);
            fillRect(c,120 + ((game_sdx/3)*2) + 120, 160 + game_sdx, 120 - ((game_sdx/3)*2), 160 - game_sdx);
        }
        if(++game_kz==30) {
            game_kz = -1;
        }
    }
```

图　8-16

```
    public void fillRect(Canvas c,int x,int y,int w,int h) {
        p.setColor(Color.BLACK);
```

```
        c.drawRect(x, y, x+w, y+h, p);
    }
    public void fillArc(Canvas c,int cx,int cy,int radio) {
        p.setColor(Color.BLACK);
        c.drawCircle(cx, cy, radio, p);
    }

    /**
     * 按键处理
     */
    public void KeyPross(){
        switch (state) {
        case LOGO:
            if(anyKeyPress){
                state = CHOOSE;
                cleanAllPress();
            }
            break;
        case CHOOSE:
                if(left_softPress){
                    music.start();
                    setPointX = 0;
                    state = MENU;
                    cleanAllPress();
                }
                if(right_softPress){
                    setPointX = 1;
                    state = MENU;
                    cleanAllPress();
                }
            break;
        case GAME_LOGO:
            if(anyKeyPress) {
                state = MENU;
                cleanAllPress();
            }
            break;
        case MENU:
            //按下一次，要等菜单条滚动完一个，所以加上一个 isMenuRoll 变量控制
            if(!isMenuRollUp){
                if(leftPress){
                    isMenuRollDown = true;
```

```
        }
    }
    if(!isMenuRollDown){
        if(rightPress){
            isMenuRollUp = true;
        }
    }
    //不需要做出滚动效果，需要确认处理写在按键处理里面
    if(!isMenuRollUp&&!isMenuRollDown){
        if(okPress) {
            switch (menuPointY) {
                case 16://开始游戏
                    startGame();
                    break;
                case 32://继续游戏
                    conGame();
                    break;
                case 48://游戏设置
                    cleanAllPress();
                    OLD_STATE = MENU;
                    state = SET;
                    break;
                case 64://游戏帮助
                    state = HELP;
                    cleanAllPress();
                    break;
                case 80://关于游戏
                    state = ABOUT;
                    cleanAllPress();
                    break;
                case 96://退出游戏
                    exit();
                    break;
            }
        }
    }
    break;
case GAME:
    if(!isGamePause){
        if(leftPress) {
            hero.moveLeft();
        }
```

```
            if(rightPress) {
                hero.moveRight();
            }
            if(upPress || okPress) {
                hero.up();
            }
        }
        break;
    case SET:
        //因为只是简单的按键处理，直接写在这里
        break;
    case HELP:
        break;
    case ABOUT:
            if(right_softPress){
                state = MENU;
                cleanAllPress();
            }
        break;

    case HERO_DEAD:

        break;
    case GAME_OVER:
        if(right_softPress) {
            state = MENU;
            cleanAllPress();
        }
        break;
    }
}

/**
 * MENU 逻辑
 */
public void menuLogic(){
    //菜单的滚动逻辑。在没有滚动完一格的时候，不释放滚动，使滚动继续
    if(isMenuRollUp){
            if(menu_count < 8){
                menuPointY -= 2;
                if(menu_count%3==0){
                menuZhuanlunY --;
```

```
                if(menuZhuanlunY < 0){
                    menuZhuanlunY = 2;
                }
                }
                if(menuPointY == 0){
                    menuPointY = 96;
                }
                menu_count++;
            }else{menu_count = 0;isMenuRollUp = false;}
    }
    if(isMenuRollDown){
            if(menu_count < 8){
                menuPointY += 2;
                if(menu_count%3==0){
                menuZhuanlunY ++;
                if(menuZhuanlunY > 2){
                    menuZhuanlunY = 0;
                }
                }
                if(menuPointY == 112){
                    menuPointY = 16;
                }
                menu_count++;
            }else{menu_count=0;isMenuRollDown = false;}
    }
}

public void run(){
    startTime = SystemClock.currentThreadTimeMillis();
    while(isRun){
        Canvas c = null;
        try{
            KeyPross();  //处理按键
            logic();     //处理逻辑
            c = sh.lockCanvas();  //锁定画布
            switch(state){ //根据不同的状态进行绘制
            case LOGO:
                drawLogo(c);
                break;
            case MENU:
                clean();
                drawMenu(c);
```

```
                break;
            case GAME:
                drawGame(c);
                break;
            case CHOOSE:
                drawChoose(c);
                break;
            case SET:
                drawSet(c);
                break;
            case HELP:
                drawHelp(c);
                break;
            case ABOUT:
                drawAbout(c);
                break;
            case HERO_DEAD:
                drawLoos(c);
                break;
            case GAME_OVER:
                drawGameOver(c);
                break;
            }

            endTime = SystemClock.currentThreadTimeMillis();
            long temp_rate = 0;
            if(endTime - startTime < rate){ //控制FPS
                temp_rate = rate;
                Thread.sleep(temp_rate - (endTime - startTime));
            }
            startTime = SystemClock.currentThreadTimeMillis();
        }catch (Exception e) {
            // TODO: handle exception
        }finally{
            if(c!=null){
                //解锁画布，写在finally，无论如何都将其解锁
                sh.unlockCanvasAndPost(c);
            }
        }
    }
}
/**
```

```
     * 存储数据
     * @param life
     * @param level
     * @param maxlevel
     */
    public void saveStage(int life,int level,int maxlevel) {
        SharedPreferences settings =
context.getSharedPreferences("data", 0);
        SharedPreferences.Editor editor = settings.edit();
        editor.putInt("life", (life + 1));
        editor.putInt("level", (level + 1));
        editor.putInt("maxlevel", (maxlevel + 1));
        editor.commit();
    }
    /**
     *
     */
    public void saveStage() {
        SharedPreferences settings =
        context.getSharedPreferences("data", 0);
        if(settings.getInt("level", 0) == 0){
            SharedPreferences.Editor editor = settings.edit();
            editor.putInt("life", 3);
            editor.putInt("level", 0);
            editor.putInt("maxlevel", 0);
            editor.commit();
        }

    }
    /**
     * 读取数据
     */
    public void restoreStage() {
        SharedPreferences settings =
        context.getSharedPreferences("data", 0);
        lsLife = settings.getInt("life", 0) - 1;
        lsLevel = settings.getInt("level", 0) - 1;
        lsMaxLevel = settings.getInt("maxlevel", 0) - 1;
    }
}
```

8.5.3 Map 类

Map 类用于存储地图以及游戏中的相关数据，代码如下所示：

```
package com.ming.last;
class Map {
    //游戏地图，数据分别为[关][挡板索引][起始 x，起始 y，类型，
    //      最大长度，速度，起始长度，剩余长度，起始运动方向]
    public static int game_map[][][] = {
        {
            {270,580,3,80,1,80,40,0},
            {215,530,1,60,1,60,30,0},
            {170,480,4,70,1,70,70,0},   //NPC 在上面行走
            {0,480,4,40,1,40,40,0},     //弹簧在这里
            {160,430,2,60,1,60,30,0},   //向右跳，左边 x
            {235,380,3,70,1,70,30,0},
            {45,400,2,40,1,40,5,0},
            {250,330,2,70,1,70,40,0},
            {150,320,1,70,1,70,30,0},
            {0,290,4,50,0,50,50,0},
            {50,240,3,50,1,50,10,0},
            {130,190,4,70,0,70,70,0},   //NPC 走
            {215,220,2,60,1,60,30,0},
            {300,190,3,60,1,60,30,0},
            {310,140,2,60,1,60,0,0},
            {193,90,4,70,0,70,70,0},    //NPC 移动
            {110,60,2,70,1,70,30,0},
            {0,40,4,70,0,70,70,0},      //结束
            {0,637,4,637,3,637,637,0}, //地板
        },
        {
            {30,580,4,60,0,60,60,0},
            {120,530,1,60,1,40,30,0},
            {205,480,2,60,1,60,30,0},
            {310,530,4,50,0,50,50,0},//  4
            {285,320,3,75,1,75,45,0},
            {175,290,2,75,1,75,45,0},
            {70,290,1,70,1,70,30,0},
            {0,240,1,60,1,60,30,0},
            {90,190,2,60,1,60,20,0},
            {160,140,1,50,1,50,30,0},
            {230,90,3,65,2,65,40,0},
            {150,40,4,50,0,50,50,0},
```

```
        {0,637,4,637,3,637,637,0}, //地板
    },
    {
        {25,580,2,60,1,60,20,0},
        {95,345,6,560,1,70,344,0}, //NPC2
        {15,190,6,400,2,70,189,0},
        {0,125,4,50,0,50,50,0},      //弹簧
        {65,37,2,60,1,60,10,0},
        {155,80,4,50,0,50,50,0},
        {275,580,2,60,1,60,20,0},
        {195,345,6,560,1,70,344,0},
        {295,190,6,400,2,70,189,0},//NPC
        {310,125,4,50,0,50,50,0},   //弹簧
        {235,37,2,60,1,60,10,0},
        {159,140,4,45,0,45,45,0},//OVER
           {0,637,4,637,3,637,637,0},     //地板
    },
    {
        {270,580,3,80,1,80,40,0},
        {215,530,1,60,1,60,30,0},
        {170,480,4,70,1,70,70,0},//NPC 在上面行走
        {0,480,4,40,1,40,40,0}, //弹簧在这
        {160,430,2,60,1,60,30,0},//向右跳，左边 x
        {235,380,3,70,1,70,30,0},
        {45,400,2,40,1,40,5,0},
        {250,330,2,70,1,70,40,0},
        {150,320,1,70,1,70,30,0},
        {0,290,4,50,0,50,50,0},
        {50,240,3,50,1,50,10,0},
        {130,190,4,70,0,70,70,0},//NPC1
        {215,220,2,60,1,60,30,0},
        {300,190,3,60,1,60,30,0},
        {310,140,2,60,1,60,0,0},
        {193,90,4,70,0,70,-10,0}, //NPC1
        {110,60,2,70,1,70,30,0},
        {0,50,4,70,0,70,70,0}, //结束
        {0,637,4,637,3,637,637,0},     //地板
    },
    {
        {25,580,2,60,1,60,30,0},
        {25,530,4,80,1,80,80,0},
        {120,350,6,500,2,30,349,0}, //向上
```

```
        {190,400,5,275,2,30,189,0},//向右
        {290,450,2,35,1,35,0,0},
        {320,520,4,40,0,40,40,0},  //弹簧
        {300,310,2,60,1,60,30,0},
        {290,260,4,70,0,70,70,0},
        {215,210,1,60,1,30,30,0},
        {55,160,5,210,2,40,54,0},//向右
        {45,110,2,60,1,40,30,0},
        {105,60,5,245,2,75,104,0},//R
        {310,40,4,50,0,50,50,0},//OVER
        {0,637,4,637,3,637,637,0}    //地板
    },
    {
        {70,580,1,60,2,60,20,0},
        {230,580,3,60,2,60,20,0},
        {150,530,4,80,0,80,80,0},//  NPC1
        {70,480,2,60,1,60,20,0},
        {230,480,2,60,1,60,20,0},
        {0,430,4,40,0,40,40,0},//弹簧
        {320,430,3,40,1,40,1,0},
        {310,370,4,50,0,50,50,0},//弹簧
        {240,370,4,30,1,0,0,0},
        {150,400,4,80,0,80,80,0},//NPC
        {85,280,2,90,1,90,60,0},
        {200,240,3,60,1,60,30,0},
        {290,210,4,70,0,70,70,0},//NPC
        {300,160,3,60,1,60,30,0},
        {215,125,1,60,1,60,30,0},
        {95,125,4,70,0,70,70,0},//NPC
        {40,70,1,65,1,65,20,0},
        {155,40,4,60,0,60,60,0},
        {0,637,4,637,3,637,637,0},    //地板
    },
    {
        {70,580,1,60,1,60,20,0},
        {160,530,2,60,1,60,40,0},
        {260,530,1,50,1,50,20,0},
        {100,480,4,65,0,65,65,0},//NPC2
        {35,435,2,60,1,60,30,0},
        {145,400,4,50,0,50,50,0},//弹簧1
        {260,400,4,50,0,50,50,0},//弹簧2
        {150,200,4,50,0,50,50,0},
```

```
        {260,225,4,50,0,50,50,0},
        {60,160,2,70,1,70,30,0},//NPC  移动缓慢的NPC
        {60,105,3,60,2,60,-10,0},//NPC
        {60,55,1,70,2,50,10,0},//NPC
        {190,55,4,65,2,65,65,0}, //NPC
        {305,40,4,55,0,55,55,0}, //结束
        {0,637,4,637,3,637,637,0},    //地板
    },
    {
        {0,637,4,637,3,637,637,0},    //地板
        {10,580,1,60,1,60,10,0},
    {50,530,2,70,2,70,-10,0},
    {90,480,2,70,2,70,-30,0},
    {130,430,2,70,2,70,-20,0},
    {160,380,1,60,2,60,-1,0},
    {190,330,2,50,1,50,20,0},
    {70,330,2,70,1,70,40,0},
    {275,330,3,60,2,60,30,0},
    {0,300,1,30,2,30,-10,0},
    {0,250,4,40,1,40,40,0},  //保护道具
    {300,270,3,50,1,50,20,0},
    {235,230,1,70,1,70,40,0},
    {140,200,2,40,1,40,-10,0},
    {210,170,2,40,1,40,-20,0},
    {140,140,2,40,1,40,-30,0},
    {55,140,1,50,1,50,30,0},
    {245,580,4,70,0,70,70,0},//NPC
    {320,545,3,40,1,40,0,0},
    {320,495,4,40,1,40,40,0},//弹簧道具
    {45,80,2,50,1,50,30,0},
    {130,60,2,50,1,50,20,0},
    {220,40,2,50,1,50,25,0},
    {300,40,4,60,1,60,60,0},
    },
    {
        {90,580,3,60,1,60,20,0},
        {190,580,1,60,1,60,20,0},
        {20,530,2,60,1,60,30,0},
        {260,530,3,70,1,70,30,0},
        {90,480,3,60,1,60,30,0},
        {190,480,1,60,1,60,30,0},
        {92,430,2,40,2,40,-10,0},
```

```
        {190,430,4,70,0,70,70,0},   //NPC1
        {128,380,2,60,1,60,20,0},
        {65,360,2,30,1,30,-10,0},   //修正一下
        {0,340,4,40,0,40,40,0},     //弹簧
        {205,330,3,60,1,60,20,0},
        {260,280,2,70,1,70,25,0},
        {170,230,4,80,0,80,80,0},   //NPC1
        {80,180,2,50,1,60,10,0},
        {0,180,4,40,0,40,40,0},
        {170,170,3,60,1,60,20,0},
        {260,85,4,70,0,70,70,0},    //NPC1
        {270,145,2,50,1,50,30,0},
        {140,55,4,80,0,80,80,0},    // NPC1
        {0,35,4,90,0,90,90,0},      //OVER
        {0,637,4,637,3,637,637,0}   //地板

    },
    {
        {25,590,1,60,2,25,25,1},
        {125,550,3,60,1,20,20,1},
        {0,510,1,70,1,30,25,0},
        {140,470,2,50,1,-20,-10,1},
        {245,550,3,25,1,5,15,1},
        {210,580,2,20,1,-50,-30,1},//隐藏块
        {255,500,2,40,1,40,15,0},
        {310,540,3,50,1,50,20,0},   //弹簧道具
        {345,440,3,50,1,0,20,1},
        {240,390,2,60,1,0,30,1},
        {130,360,1,60,1,2,30,1},
        {50,305,1,50,1,50,20,0},
        {95,250,2,100,1,30,30,1},   //加怪挡板
        {0,200,1,40,1,40,20,0},     //护甲道具
        {150,200,3,90,1,90,40,0},   //加怪挡板
        {250,230,4,20,1,30,20,1},
        {320,170,3,50,1,20,20,0},
        {340,330,4,20,1,20,20,0},   //加生命道具
        {250,120,1,85,1,85,25,0},   //加怪挡板
        {200,100,4,20,1,20,20,1},
        {90,120,2,100,1,40,40,1},   //加怪挡板
        {70,70,2,30,2,30,-10,0},
        {0,35,4,40,1,40,10,1},      //过关
        {0,637,4,637,3,637,637,0}   //地板
```

```
},
    {
        {310,600,3,50,1,50,30,0},
        {230,550,1,70,1,20,20,1},       //加怪挡板
        {275,500,2,50,1,50,20,0},
        {230,440,4,30,1,30,30,0},
        {120,380,3,100,1,100,30,0},     //加怪挡板
        {90,440,4,0,1,20,0,0},
        {70,505,2,90,1,40,40,1},        //加怪挡板
        {0,535,4,0,1,20,0,0},           //加生命
        {0,465,4,0,1,20,0,0},           //弹簧
        {0,300,4,0,1,30,0,0},
        {50,240,2,100,1,100,30,0},
        {190,180,3,60,1,20,30,1},
        {140,120,4,0,1,20,0,0},         //加护甲道具
        {225,200,3,80,1,80,30,0},       //加怪挡板
        {280,140,3,70,1,70,40,0},       //加怪挡板
        {240,85,4,0,1,20,0,0},
        {320,35,3,70,1,40,40,1},        //门
        {0,637,4,637,3,637,637,0}       //地板
    },
    {
         {150,590,3,50,1,50,32,0},
        {90,535,4,90,1,90,30,0},        //加怪挡板
        {50,465,2,100,1,100,30,1},      //加怪挡板
        {0,430,4,0,0,20,0,0},
        {50,380,5,210,2,30,40,0},       //横移
        {275,360,6,550,2,30,359,0},     //纵移
        {340,450,4,0,0,20,0,0},         //弹簧
        {340,530,4,0,0,20,0,0},         //加生命
        {340,270,4,0,0,30,0,0},
        {270,230,6,250,-2,30,120,0},
        {210,120,4,0,0,30,0,0},
        {120,180,1,90,1,80,35,0},       //加怪挡板
        {50,190,4,0,0,20,0,0},
        {0,140,4,0,0,30,0,0},
        {40,80,2,100,1,80,40,0},        //加怪挡板
        {0,40,4,0,0,20,0,0},            //加速
        {90,40,5,185,2,40,49,0},        //横移
        {300,35,4,60,3,60,640,0},       //门
        {0,637,4,637,3,637,637,0},      //地板
```

```
    },
    {
        {0,340,4,30,1,30,30,0},
        {100,380,4,100,1,100,35,0},     //加怪挡板
        {0,520,1,60,1,20,20,1},         //护甲
        {80,430,1,90,1,90,30,0},        //加怪挡板
        {110,520,4,90,1,90,30,0},       //加怪挡板
        {200,555,5,300,2,30,199,1},     //横移
        {330,510,4,0,0,30,0,0},         //弹簧
        {330,320,4,0,0,30,0,0},
        {280,270,2,30,1,30,-11,0},
        {320,95,6,230,2,40,90,0},       //纵移
        {200,80,4,100,1,100,30,0},      //加怪
        {170,120,2,30,1,0,-15,1},
        {205,170,4,0,0,20,0,0},         //定时
        {145,210,2,85,1,85,20,0},       //加怪
        {75,260,4,80,1,80,30,1},        //加怪
        {10,230,6,231,-2,30,80,1},      //纵移
        {0,35,4,0,0,40,0,0},            //门
        {0,637,4,637,3,637,637,0}       //地板
    },
    {
        {0,35,4,0,0,60,0,0},            //门
        {130,90,4,0,0,100,0,0},
        {110,125,2,100,1,100,50,0},     //怪
        {50,160,4,0,0,30,0,0},
        {50,210,1,100,1,100,50,0},      //怪
        {170,260,4,50,1,50,30,1},       //生命
        {70,310,4,0,0,100,0,0},         //怪
        {0,320,6,520,2,40,319,0},       //纵移
        {330,600,4,0,0,50,0,0},         //起始
        {200,550,2,100,1,100,50,0},     //怪
        {300,500,3,60,1,60,20,0},       //弹簧
        {320,320,4,0,0,40,0,0},         //护甲
        {200,440,2,100,1,100,50,0},     //怪
        {120,410,2,80,1,80,40,0},       //怪
        {50,480,4,0,0,90,0,0},          //怪
        {260,70,4,0,0,30,0,0},
        {320,35,4,0,0,40,0,0},          //加速
       {0,637,4,637,3,637,637,0}        //地板
    },
    {
```

```
        {75,570,1,80,2,80,60,0},
        {165,520,2,80,1,80,40,0},
        {140,470,1,50,2,50,30,0},
        {65,420,1,70,1,70,40,0},
        {220,420,3,70,1,70,50,0},
        {0,370,4,50,0,50,50,0},
        {280,370,3,50,1,50,30,0},
        {25,320,4,40,0,40,40,0},
        {190,320,2,80,1,80,50,0},
        {160,270,1,80,1,80,50,0},
        {255,220,2,70,1,70,30,0},
        {165,170,1,70,1,70,40,0},
        {76,170,1,60,1,60,30,0},
        {25,120,2,50,1,50,20,0},
        {115,70,4,30,1,30,30,0},
        {180,70,4,35,0,35,35,0},
        {310,70,4,50,0,50,50,0},
        {0,637,4,637,3,637,637,0},     //地板
    },
};
//怪物数组，数据分别为[关][怪物索引][x,y,类型，怪物所在挡板的索引，运动方向，速度]
public static int npc_array[][][] ={
    {
        {170,465,1,2,1,1},
        {130,175,3,11,1,1},
        {193,75,2,15,1,1},
    },
    {

    },
    {
        {105,331,1,1,1,2},
        {210,331,1,7,0,2},
    },
    {
        {170,465,1,2,1,2},
        {130,175,1,11,1,3},
        {193,75,1,15,1,2}
    },
    {
        {25,515,1,1,0,1},
```

```
        {130,45,2,11,0,2},
        {290,245,3,7,0,1},
    },
    {
        {150,515,1,2,1,2},
        {150,385,1,9,1,1},
        {310,195,1,12,0,1},
        {115,110,1,15,0,1},
        {95,265,2,10,1,1}
    },
    {
        {100,465,2,3,0,1},
        {190,40,1,12,1,1}
    },
    {
        {245,565,1,17,1,2}
    },
    {
        {200,415,1,7,1,1},
        {185,215,2,13,1,2},
        {300,70,1,17,0,1},
        {200,40,3,19,0,3}
    },
    {
        {100,235,1,12,0,2},
          {200,185,2,14,0,2},
          {250,105,2,18,0,2},
          {110,105,3,20,0,2}
    },
    {
        {230,535,1,1,0,2},
        {140,365,1,4,1,2},
        {90,490,2,6,0,2},
        {70,225,3,10,1,2},
        {210,185,2,13,1,2},
        {320,125,3,14,0,2}
    },
    {
        {160,520,3,1,0,2},
        {45,450,2,2,1,2},
        {180,165,1,11,0,2},
        {180,65,1,14,1,2}
```

```
    },
    {
    {100,365,1,1,0,2},
    {150,415,2,3,0,2},
    {120,505,3,4,1,2},
    {250,65,3,10,1,2},
    {200,195,2,13,1,2},
    {80,245,1,14,1,2}
    },
    {
        {160,110,1,2,1,2},
    {100,195,2,4,1,2},
    {70,295,3,6,1,2},
    {250,535,3,9,1,2},
    {250,425,1,12,1,2},
    {160,395,2,13,1,2},
    {50,465,2,14,1,2}
    },
    {
        {0,355,1,5,1,2},
    },
};
//道具数组，数据分别为 [关][道具索引][x,y,类型]
public static int bonus[][][] = {
    {
        {60,388,0},
        {16,468,4},
    },
    {
        {329,518,4},
    },
    {
        {105,372,0},
        {243,442,1},
        {15,113,4},
        {333,113,4},
    },
    {
        {20,90,1},
        {15,468,4},
        {35,278,4},
    },
```

```
{
     {335,508,4},
},
{
    {230,280,0},
    {45,218,1},
    {15,418,4},
    {325,358,4},
},
{
                {15,190,0},
                {167,388,4},
                {15,390,4},
                {325,85,3},
                {330,330,4},
                {335,510,4},
},
{
    {300,85,0},
    {335,482,4},
    {15,238,4},
    },
{
     {330,25,1},
    {15,328,4},
},
{
    {345,318,0},
    {348,528,4},
    {0,188,1}
},
{
    {0,453,4},
    {0,523,0},
    {140,108,1}
},
{
    {0,418,0},
    {345,438,4},
    {345,518,0},
    {0,28,2}
},
```

```
    {
        {3,508,1},
        {340,498,4},
        {210,158,0}
    },
    {           {200,248,0},
        {345,488,4},
        {335,308,1},
        {345,23,2}
    },
    {
        {200,0,2},
        {40,180,3},
        {40,308,4},
        {348,265,0},
    },
};
//Hero 数组，数据分别为 [关][x,y,初始状态]
public static int hero_array[][]={
    {180,590,2},
    {60,590,2},
    {60,590,2},
    {60,590,2},
    {300,558,2},
    {60,590,2},
    {60,590,2},
    {40,568,2},
    {300,558,2},
    {25,558,2},
    {328,568,2},
    {168,558,2},
    {0,308,2},
    {280,605,2},
    {60,590,2},
};
//过关点数组，数据分别为[关][x,y,初始状态]
public static int guanqia[][]={
    {20,5,45},
    {175,5,45},
    {182,105,145},
    {20,15,55},
    {340,5,45},
```

```
        {185,5,45},
        {340,5,45},
        {340,5,45},
        {20,0,40},
        {20,0,40},
        {340,0,40},
        {340,0,40},
        {20,0,40},
        {20,0,40},
         {340,35,75},
    };
}
```

8.5.4 Npc 类

```
package com.ming.last;

import android.graphics.Canvas;
import android.graphics.Paint;

public class Npc {
    public int nx,ny;
    public int style;
    public int boardInRow;
    public int DIR;
    public int speed;
    public int nw=15,nh=15;
    public int nextF;
    public int nowRow;
    /**
     *
     * @param x Npc的x坐标
     * @param y Npc的y坐标
     * @param s Npc的类型(怪物的样子)
     * @param bir Npc所在木板数组的位置
     * @param d   Npc的方向：0表示左移动，1表示右移动
     * @param sp Npc的速度
     */
    public Npc(int x ,int y ,int s, int bir ,int d ,int sp){
        nx = x;
        ny = y;
        style = s - 1;
        boardInRow = bir;
```

```
        DIR = d;
        speed = sp;
    }
    /**
     * 绘制 Npc
     * @param c
     * @param p
     */
    public void drawNpc(Canvas c,Paint p){
            Tools.setClip(c,nx+Tools.offx, ny + Tools.offy, 15, 15);
           c.drawBitmap(GameView.imggameNpc, nx - nextF*15+Tools.offx,
ny + Tools.offy - nowRow*15, p);
            Tools.resetClip(c);
    }
    /**
     * 移动 Npc
     * @param allb
     */
    public void move(Board[] allb){
        int sped = allb[boardInRow].speed;
        if(DIR==0){ //根据不同的移动类型，进行不同的移动
            nowRow = style*2;
            if(allb[boardInRow].style == 5){
                nx -= speed - sped;
            }else{
                nx -= speed;
            }
            if(allb[boardInRow].style == 6){
                ny += sped;
            }
            if(nx < allb[boardInRow].bx){DIR=1;}
        }
        else{if(DIR==1){
            nowRow = style*2 + 1;
            if(allb[boardInRow].style == 5){nx += speed + sped;}
            else{nx += speed;}
            if(allb[boardInRow].style == 6){
                ny += sped;
            }
            if((nx+nw) > (allb[boardInRow].bx +
allb[boardInRow].addx)){DIR=0;}
            }
```

```
        }
        nextFrame(); //播放动画
    }
    /**
     * 播放动画
     */
    public void nextFrame(){
        nextF++;
        if(nextF > 1){nextF=0;};
    }
}
```

8.5.5 Bonus（道具）类

```
/**
     * 道具构造
     * @param bx 道具x坐标
     * @param by 道具y坐标
     * @param bKinds 道具种类
     */
    public Bonus(int bx,int by,int bKinds){
        bonusX = bx;
        bonusY = by;
        bonusKind = bKinds;
    }
```

8.5.6 Hero 类

```
package com.ming.last;
import android.content.Context;
import android.graphics.Bitmap;
import android.graphics.Canvas;
import android.graphics.Paint;

class Hero  {
    Context ct;
    public static final int STATE_STAND = 0;  //站立
    public static final int STATE_LEFT  = 2;   //左走
    public static final int STATE_RIGHT = 3;  //右走
    public static final int STATE_JUMPL = 1;  //左跳
    public static final int STATE_JUMPR = 4;  //右跳

    public final int heroW = 32,heroH = 32;
    public static int life = 0;//生命
```

```
public boolean isJump;      //是否跳跃
public boolean isJumping;  //是否跳跃中
public int state;          //hero 状态
public int hero_frame;      //hero 帧
public int hero_x;          //hero x 速度
public int hero_y;          //hero y 速度
public int hero_speed;      //hero 速度
public int hero_addSpeed;  //hero 加速后速度
public int hero_jump_speed=10;//hero 跳跃速度
public int hero_jump_addSpeed;//hero 踏到加速板后的跳跃速度
public int hero_jump_high; //hero 跳跃高度
public boolean isAddSpeed; //hero 是否踏到加速板
public boolean isPower;          //是否处于无敌状态
public boolean isPause;          //是否吃到护甲道具
public boolean isDown;
public int downdistance;//下落高度
public int countZ;//暂停时间
public int countS;//速度时间
public int countB;//保护时间
public int nextF;
public int onBonus;//用户存放踩到木板上的数据
public boolean isStandBorad;//是否踩上木板
public int[][] action = {
       {11}, //左跳
       {6,7,8,9,10},//左
       {0,1,2,3,4},//右
       {5}, //右跳
};
public Bitmap imgHero;
public Hero(){
}
/**
 * 初始化人物属性 每一关都调用此方法，初始化 Hero 属性
 * @param img
 * @param x
 * @param y
 * @param sta
 */
public void initHero(Bitmap img,int x,int y,int sta) {
    imgHero = img;
    hero_x= x;
    hero_y= y;
```

```
        state = sta;
        isAddSpeed = false;
        isPause = false;
        isPower = false;
        isDown = false;
        hero_frame = 2; //初始为站立状态
        hero_speed = 3;
        hero_addSpeed = 5;
    }

    /**
     * 绘制Hero
     * @param g
     */
    public void paint(Canvas c,Paint p) {
        Tools.setClip(c,hero_x+Tools.offx, hero_y + Tools.offy, 32, 32);
        c.drawBitmap(imgHero, hero_x-(action[state-1][hero_frame])%6 *
32+Tools.offx, hero_y-(action[state-1][hero_frame])/6*32 + Tools.offy -
GameView.tmph * 64,p);
        Tools.resetClip(c);
        if(isPower) {
            Tools.setClip(c,hero_x - 9+Tools.offx, hero_y + Tools.offy -
9, 50, 50);
            c.drawBitmap(GameView.imgPower,(hero_x - 9) -
nextF*50+Tools.offx, hero_y + Tools.offy - 9, p);
            Tools.resetClip(c);
        }
    }
    /**
     * 左移动
     */
    public void moveLeft() {
        state = STATE_LEFT;
        if(isJumping) {
            state = STATE_JUMPL;
        }
        hero_x -= hero_speed;
        if(hero_x < -4){hero_x = -4;}
        nextFrame();
    }
    /**
     * 右移动
```

```
    */
   public void moveRight() {
       state = STATE_RIGHT;
       if(isJumping) {
          state = STATE_JUMPR;
       }
       hero_x += hero_speed;
       if(hero_x + 28 > 360){hero_x = 360 - 28;}
       nextFrame();
   }

   public void up() {
       if(!isDown){
          if(!isJump&&!isJumping) {
              isJump = true;
              isJumping = true;
          }
       }
   }
   /**
    * 播放动画
    */
   public void nextFrame() {
       hero_frame = hero_frame < action[state-1].length-1 ?
++hero_frame:0;
   }
   /**
    * 逻辑
    */
   public void logic(Board[] allb) {
       if(isJump) {
          jump();
       }
       else{
          down(allb);
       }
       choiceSpeed();
       if(isPause){        //定时所有挡板与怪物 5 秒
          if(++countZ>= 100) {
              isPause = false;
              countZ = 0;
          }
```

```
        }
        if(isPower) {     //hero 无敌 5 秒
            if(++countB >= 100) {
                isPower = false;
                countB = 0;
            }
        }
        if(isAddSpeed) {  //hero 加速 8 秒
            if(++countS >= 160) {
                isAddSpeed = false;
                countS = 0;
            }
        }
        if(++nextF > 5){nextF=0;}

    }
    /**
     * 跳跃上升
     */
    public void jump() {
        hero_frame = 0;
        if(state == STATE_LEFT) {state = STATE_JUMPL;}
        if(state == STATE_RIGHT) {state = STATE_JUMPR;}
        hero_y-= hero_jump_speed;
        hero_jump_speed --;
        if(hero_jump_speed <= 0){
            isJump = false;
        }
    }
    /**
     * 下降
     */
    public void down(Board[] allb){
        isDown = true;
        if(hero_jump_speed < 10){
            hero_jump_speed++;
        }
        hero_y += hero_jump_speed;
        downdistance += hero_jump_speed;
        Tools.checkHeroAndBoard(this, allb); //检测是否站在了挡板上

        if(isStandBorad){//如果是刚改变 hero 状态为非跳跃状态，修正 hero 坐标
```

```
            hero_y = onBonus - 32;
            isJumping = false;
            hero_jump_speed = 11;
            if(state == STATE_JUMPL) {state = STATE_LEFT;}
            if(state == STATE_JUMPR) {state = STATE_RIGHT;}
            isDown = false;
            if(!isPower){ //如果不在保护状态下，并且下降距离超过300，则掉一命
                if(downdistance > 300){
                    life--;
                    countB = 0;
                    isPower = true;
                }
            }
            downdistance = 0;
        }
    }
    /**
     * Hero 速度判断，即是否为加速状态
     */
    public void choiceSpeed() {
        if(isAddSpeed) {
            hero_speed = hero_addSpeed;
        }else {
            hero_speed = 3;
        }
    }
}
```

8.5.7 Tools（工具）类

```
//工具类Tools 主要封装了一些通用方法，以及挡板、道具、Npc的绘制和逻辑
package com.ming.last;

import android.content.Context;
import android.content.res.Resources;
import android.graphics.Bitmap;
import android.graphics.BitmapFactory;
import android.graphics.Canvas;
import android.graphics.Color;
import android.graphics.Paint;
import android.graphics.Rect;
import android.graphics.Region;
import android.graphics.Paint.Style;
```

```
public class Tools {
    Context ct;
    public static int offx = 120,offy = -320;//x 轴、y 轴方向的滚屏
    public static int screenH = 320,screenW = 240;//屏幕宽高
    //构造方法
    public Tools(Context contest){
        ct = contest;
    }
    /**
     * 通过 ID 获得图片对象
     * @param id
     * @return
     */
    public Bitmap createBmp(int id){
        Bitmap bmp = null;
        Resources res = null;
        res = ct.getResources();
        bmp = BitmapFactory.decodeResource(res, id);
        return bmp;
    }
    /**
     * 通过数组获得图片数组
     * @param id 数组
     * @param num 数组长度
     * @return
     */
    public Bitmap[] createBmp(final int id[],int num){
        Bitmap bmp[] = null;
        bmp = new Bitmap[num];
        for (int i = 0; i < bmp.length; i++) {
            Resources res = null;
            res = ct.getResources();
            bmp[i] = BitmapFactory.decodeResource(res, id[i]);
        }
        return bmp;
    }
    /**
     * 检测碰撞
     * @param x 触点的 x 坐标
     * @param y 触点的 y 坐标
     * @param r    矩形对象
```

```
 * @return
 */
public boolean checkRectTouch(int x,int y,Rect r){
    if(r.contains(x, y)){
        return true;
    }
    return false;
}
/**
 * 滚屏逻辑
 * @param hero 人物对象
 */
public static void setCamera(Hero hero){
    offy = -(hero.hero_y - ((screenH/5)*3));
    if(offy <= -screenH){offy = -screenH;}
    if(offy > 0){
        offy = 0;
    }
    offx = 120 - (hero.hero_x - (screenW>>1));
    if(offx <= 0){offx = 0;}
    if(offx > 120){
        offx = 120;
    }
}
/**c.save()与c.restore()必须成对出现**
/**
 * 设定剪裁区
 * @param c
 * @param x
 * @param y
 * @param width
 * @param height
 */
public static void setClip(Canvas c,int x,int y,int width,int height)
{
    c.save();
    c.clipRect(x, y, x+width, y+height,Region.Op.REPLACE);
}
/**
 * 恢复剪裁区
 * @param c
 */
```

```
    public static void resetClip(Canvas c) {
        c.restore();
    }
    /**
     * 清屏
     * @param c
     * @param p
     */
    public static void cleanScreen(Canvas c,Paint p) {
        p.setColor(Color.WHITE);
        p.setStyle(Style.FILL);
        c.drawRect(new Rect(0,0,360,480), p);
    }
    /*****************************************************/
    /**
     * 绘制所有木板
     * @param allb
     * @param g
     */
    public static void drawAllBoard(Canvas c,Board[] allb,Paint p){
        for (int i = 0; i < allb.length; i++) {
            if((allb[i].bx + Tools.offx + allb[i].addx) > 120
                    && allb[i].bx + Tools.offx < 360
                        && (allb[i].by + Tools.offy) < 330
                            && (allb[i].by + Tools.offy + 15) > -10){
                if(allb[i].style == 2){
                    Tools.setClip(c,allb[i].bx+Tools.offx, allb[i].by+
Tools.offy, allb[i].addx, 14);
                    c.drawBitmap(GameView.imggameBoard,
allb[i].bxC+Tools.offx, allb[i].by + Tools.offy - GameView.tmpT*20, p);
                    Tools.resetClip(c);

                }
                else if(allb[i].style == 3){
                    Tools.setClip(c,allb[i].bx+Tools.offx, allb[i].by+
Tools.offy, allb[i].addx, 14);
                    c.drawBitmap(GameView.imggameBoard,
allb[i].bxR+Tools.offx, allb[i].by + Tools.offy - GameView.tmpT*20, p);
                    Tools.resetClip(c);
                }
                else{
                    Tools.setClip(c,allb[i].bx+Tools.offx, allb[i].by+
```

```
Tools.offy, allb[i].addx, 14);
                c.drawBitmap(GameView.imggameBoard,
allb[i].bx+Tools.offx, allb[i].by + Tools.offy - GameView.tmpT*20, p);
                Tools.resetClip(c);
            }
            if(allb[i].style == 4 && allb[i].speed == 3){
                Tools.setClip(c,allb[i].bx    +    120+Tools.offx,
allb[i].by + Tools.offy, allb[i].addx, 14);
                c.drawBitmap(GameView.imggameBoard,  allb[i].bx  +
120+Tools.offx, allb[i].by + Tools.offy - GameView.tmpT*20, p);
                Tools.resetClip(c);
                Tools.setClip(c,allb[i].bx    +    240+Tools.offx,
allb[i].by + Tools.offy, allb[i].addx, 14);
                c.drawBitmap(GameView.imggameBoard,  allb[i].bx  +
240+Tools.offx, allb[i].by + Tools.offy - GameView.tmpT*20, p);
                Tools.resetClip(c);
            }
        }
    }
  }
  /**
   * 检测人物与木板的碰撞
   * @param h
   * @param allb
   */
  public static void checkHeroAndBoard(Hero h,Board[] allb){
      for(int i = 0 ; i < allb.length ; i++) {
          if(allb[i].addx > 0){
          if(allb[i].style == 6){
              if((allb[i].by - h.hero_y <= 32) &&
                      (allb[i].by - h.hero_y >= 19)) {
                  if((allb[i].bx < (h.hero_x + 24)) &&
                          ((allb[i].bx+allb[i].addx)>(h.hero_x+8)))
{
                      h.hero_y += allb[i].speed;
                      h.onBonus = allb[i].by;
                      h.isStandBorad = true;
                      break;
                  }
              }
          }else{
              if((allb[i].by - h.hero_y <= 32) &&
```

```
                (allb[i].by - h.hero_y >= 21)) {
            if((allb[i].bx < (h.hero_x + 24)) &&
                ((allb[i].bx+allb[i].addx)>(h.hero_x+8)))
{
                if(allb[i].style == 5){
                    h.hero_x += allb[i].speed;
                    }
                h.onBonus = allb[i].by;
                h.isStandBorad = true;
                break;
            }
        }
        }
    }
    h.isStandBorad = false;
  }
}

/**
 * 让所有的模板移动
 * @param allb
 */
public static void moveAllBoard(Board[] allb){
    for (int i = 0; i < allb.length; i++) {
        if (allb[i].style == 1) {
            allb[i].changeByLeft();
        } else {
            if (allb[i].style == 2) {
                allb[i].changeByCenter();
            } else {
                if (allb[i].style == 3) {
                    allb[i].changeByRight();
                } else if (allb[i].style == 5) {
                    allb[i].moveLeftAndRight();
                } else if (allb[i].style == 6) {
                    allb[i].moveUpAndDown();
                }
            }
        }
    }
}
/**
```

```
 * 检测人物和Npc的碰撞
 * @param h    人物对象
 * @param allnNpc对象数组
 */
public static void checkHeroAndNpc(Hero h , Npc[] alln){
    if(!h.isPower){
        for (int i = 0; i < alln.length; i++) {
            if((h.hero_x + 32) > alln[i].nx
                &&
                h.hero_x < (alln[i].nx + alln[i].nw)
                &&
                (h.hero_y + 32) > alln[i].ny
                &&
                h.hero_y < (alln[i].ny + alln[i].nh)) {

                h.hero_x -= 5;
                h.hero_y -= 5;
                h.countB = 0;
                h.isPower = true;
                Hero.life--;
            }
        }
    }
}
/**
 * 判断人物与道具碰撞的方法
 * @param h
 * @return
 */
public static void withBonusCollision(Hero h,Bonus[] b) {
    for (int i = 0; i < b.length; i++) {
        if(b[i].bonusKind != 4 && b[i].isShow) {
            if ((h.hero_x + h.heroW) < b[i].bonusX
                    || (h.hero_y + h.heroH) < b[i].bonusY
                        || (b[i].bonusX + b[i].BW) < h.hero_x
                            || (b[i].bonusY + b[i].BH) < h.hero_y) {

            }else {
                if(b[i].bonusKind == Bonus.BS_CAP) {
                    h.countB = 0;
                    h.isPower = true;
                    b[i].isShow = false;
```

```
                }else if(b[i].bonusKind == Bonus.BS_TIGER) {
                    if(Hero.life < 8) {
                        Hero.life++;
                    }
                    b[i].isShow = false;
                }else if(b[i].bonusKind == Bonus.BS_TIMER){
                    h.countZ = 0;
                    h.isPause = true;
                    b[i].isShow = false;
                }else if(b[i].bonusKind == Bonus.BS_SPEED) {
                    h.countS = 0;
                    h.isAddSpeed = true;
                    b[i].isShow  = false;
                }
            }
        }
        if(b[i].bonusKind == 4) {
            if(h.isDown) {
                if (b[i].bonusY - (h.hero_y) <= 32 && b[i].bonusY -
(h.hero_y) >= 21) {
                    if((b[i].bonusX < (h.hero_x + 30))
                            && ((b[i].bonusX + 12) > (h.hero_x + 12)))
{
                        h.isJump = true;
                        h.isJumping = true;
                        h.downdistance = 0;
                        h.hero_jump_speed = 20;
                    }
                }
            }
        }
    }
}
/**
 * 移动所有的Npc
 * @param alln  Npc数组
 * @param allb  木板数组
 */
public static void moveAllNpc(Npc[] alln , Board[] allb){
    for (int i = 0; i < alln.length; i++) {
        alln[i].move(allb);
    }
```

```
    }
    /**
     * 绘制所有的 Npc
     * @param g
     * @param alln
     */
    public static void drawAllNpc(Canvas c,Npc[] alln,Paint p){
        for (int i = 0; i < alln.length; i++) {
            if((alln[i].nx + Tools.offx + alln[i].nw) > 120
                    && (alln[i].nx + Tools.offx) < 360
                        && (alln[i].ny + Tools.offy) < 330
                            && (alln[i].ny + Tools.offy + alln[i].nh) >
-10){
                alln[i].drawNpc(c,p);
            }
        }
    }
    /**
     * 绘制道具
     * @param g
     */
    public static void drawAllBonus(Canvas c,Bonus allBonus[],Paint p)
{
        for(int i = 0;i < allBonus.length;i++) {
            if((allBonus[i].bonusX + Tools.offx + allBonus[i].BW) > 120
                    && (allBonus[i].bonusX + Tools.offx) < 360
                        && (allBonus[i].bonusY + Tools.offy) < 330
                            &&   (allBonus[i].bonusY   +   Tools.offy   +
allBonus[i].BH) > -10){
                if(allBonus[i].isShow) {
                    Tools.setClip(c,allBonus[i].bonusX+Tools.offx,
allBonus[i].bonusY + Tools.offy, 12, 12);
                    c.drawBitmap(GameView.imgGameDaoju,
allBonus[i].bonusX    -    allBonus[i].bonusKind*    12+Tools.offx,
allBonus[i].bonusY+Tools.offy,p);
                    Tools.resetClip(c);
                }
            }
        }
    }
    /**
     * 帮助文字
```

```
 */
public static final String[] strHelp = {
    "            帮助",
    "",
    "移动：触屏左右按钮",
    "",
    "跳跃：触屏跳跃按钮",
    "",
    "玩家通过控制主角通",
    "",
    "过重重障碍，到达通",
    "",
    "关之门。",
    "",
    "         道具说明",
    "",
    "拾取生命，主角生命",
    "",
    "值加 1",
    "",
    "拾取加速，在 8 秒内",
    "",
    "",
    "",
    "主角移动速度增加",
    "",
    "拾取护甲，主角在 5",
    "",
    "秒内碰到怪物不会",
    "",
    "死亡",
    "",
    "拾取定时，所有的挡",
    "",
    "板与怪物停止移动 5 秒",
    "",
    "踩到弹簧，主角跳跃",
    "",
    "高度增加"
};
/**
 * 关于文字
```

```
    */
    public static final String[] strAbout= {
        "知了工作室出品",
        "",
        "邮箱：",
        "",
        "duskybird",
        "",
        "@yeah.net",
        "",
        "感谢您使用本产品"
    };
}
```

8.5.8　Music 类

```
package com.ming.last;

import android.content.Context;
import android.media.MediaPlayer;

public class Music {
    public MediaPlayer player;//Media 对象
    public Context context;
    /**
     * 构造
     * @param con
     */
    public Music(Context con){
        context = con;
        play();
    }
    /**
     * 加载音乐资源
     */
    public void play() {
        if (player == null) {
            player = MediaPlayer.create(context, R.raw.music);
            player.setLooping(true);
        }
    }
    /**
     * 播放音乐
     */
```

```
    public void start(){
        if(player!=null)
        {player.start();}
    }
    /**
     * 暂停音乐
     */
    public void pause(){
        if(player!=null&&player.isPlaying())
        {player.pause();}
    }
    /**
     * 中止音乐
     */
    public void stop(){
        if(player!=null&&player.isPlaying()){
            player.stop();
        }
    }
}
```

8.5.9 AndroidManifest.xml 文件

```
 <?xml version="1.0" encoding="utf-8"?>
<manifest  xmlns:android="http://schemas.android.com/apk/res/android"
   package="com.ming.last"                    android:versionCode="1"
   android:versionName="1.0">
<application                         android:icon="@drawable/icon"
   android:label="@string/app_name">
<activity                           android:name=".GameActivity"
   android:label="@string/app_name">
    <intent-filter>
      <action android:name="android.intent.action.MAIN" />
      <category android:name="android.intent.category.LAUNCHER" />
    </intent-filter>
</activity>
</application>
<uses-sdk android:minSdkVersion="3" />
</manifest>
```

8.5.10 string.xml 文件

```
 <?xml version="1.0" encoding="utf-8"?>
<resources>
<string name="hello">Hello World, GameActivity!</string>
<string name="app_name">TanBoHu</string>
</resources>
```

参考文献

[1]（美）Sierra K 著. Head First Java（中文版）.第 2 版.Tacwaw 公司译. 北京：中国电力出版社，2007.

[2] 杨丰盛编著. Android 应用开发揭秘. 北京：机械工业出版社，2010.

[3]（美）Vladimir Silva 著.精通 Android 游戏开发. 王恒，苏金同译. 北京：人民邮电出版社，2011.

[4] 李刚著. 疯狂 Android 讲义.第 3 版. 北京：电子工业出版社，2015.